WITHDRAWN

PERGAMON INTERNATIONAL LIBRARY
of Science, Technology, Engineering and Social Studies
The 1000-volume original paperback library in aid of education,
industrial training and the enjoyment of leisure
Publisher: Robert Maxwell, M.C.

Radioactivity and Its Measurement

SECOND EDITION (SI UNITS)

Revised and Expanded

Some Other Pergamon Titles of Interest

BOWLER, M. G.:
Nuclear Physics

CHOPPIN, G. & RYDBERG J.:
Nuclear Chemistry: Theory and Applications

GIBSON, W. M.:
The Physics of Nuclear Reactions

MARCH, N.:
Self-Consistent Fields in Atoms

THEWLIS, J.:
Concise Dictionary of Physics and Related Subjects 2nd Edition

Radioactivity and Its Measurement

by

W. B. MANN
Chief, Radioactivity Section,
National Bureau of Standards

R. L. AYRES
Radiochemist, National Bureau of Standards

and

S. B. GARFINKEL†
Physicist, National Bureau of Standards

SECOND EDITION (SI UNITS)

Revised and Expanded

PERGAMON PRESS

Oxford · New York · Toronto · Sydney · Paris · Frankfurt

U.K.	Pergamon Press Ltd., Headington Hill Hall, Oxford OX3 0BW, England
U.S.A.	Pergamon Press Inc., Maxwell House, Fairview Park, Elmsford, New York 10523, U.S.A.
CANADA	Pergamon of Canada, Suite 104, 150 Consumers Road, Willowdale, Ontario M2J 1P9, Canada
AUSTRALIA	Pergamon Press (Aust.) Pty. Ltd., P.O. Box 544, Potts Point, N.S.W. 2011, Australia
FRANCE	Pergamon Press SARL, 24 rue des Ecoles, 75240 Paris, Cedex 05, France
FEDERAL REPUBLIC OF GERMANY	Pergamon Press GmbH, 6242 Kronberg-Taunus, Hammerweg 6, Federal Republic of Germany

First edition 1966

Second edition 1980

British Library Cataloguing in Publication Data

Mann, Wilfrid Basil
Radioactivity and its measurement.—2nd ed.—
(Pergamon international library).
1. Radioactivity—I. Title—II. Ayres, R. L.—III.
Garfinkel, Samuel Bernard
539.7'5   QC795   79-40881

ISBN 0-08-025028-9 hardcover
ISBN 0-08-025027-0 flexicover

The first edition was published by D. Van Nostrand Company Inc. under the title *Radioactivity and Its Measurement*.

Printed in Great Britain by A. Wheaton & Co. Ltd., Exeter

Contents

Preface

THIS book is a revision of *Radioactivity and Its Measurement* by W. B. Mann and S. B. Garfinkel, published in 1966. That part of the first edition which was mostly written by our late friend and colleague, Sam Garfinkel, namely Chapters 6, 7 and 8, with the exception of a few figures, has been rewritten because the nucleonic measuring techniques applicable in 1966 are no longer valid in 1979, such has been the rate of progress in solid-state instrumentation in the fifteen years that have elapsed since the manuscript of the 1966 edition was written. His name is retained as an author, both in his memory and as a token of respect for his many contributions to the measurement of radioactivity.

The first five chapters, dealing with the historical development of radioactivity between its discovery in 1896 to the period just before World War II, are essentially the same as in the first edition, with some revision in the light of new knowledge and the replacement of old units by those of the Système International (SI). There have also been a few corrections, the most unfortunate of which related to the energetics of alpha-particle decay, when in changing from one method of derivation to the present one, twice the mass of the electron was inadvertently omitted from the typescript, thereby giving rise to a page that was deficient by 1.022 MeV.

Again, the present authors have revised this book in a personal capacity and it carries no *imprimatur* from their place of employment, the National Bureau of Standards. It does, however, because of their association, and that of the Bureau, with the preparation of the National Council on Radiation Protection and Measurements (NCRP) Report No. 58, *A Handbook of Radioactivity Measurements Procedures*, reflect to some extent what is *not* in that Report. Extensive reference has been made to NCRP Report 58, but there is little duplication. Much of the text prepared for this book on gas ionization detectors was adapted for NCRP Report 58, but on the subjects of electronic instrumentation, solid-state detectors, the statistics of counting, and, for example, the Fano factor, this book supplements NCRP Report 58, and either leads up to, or extends, the treatment given in the Report.

The tremendous increase in the last three decades in the availability of radioactive isotopes of every known element in nature has, in turn, given rise to an enormous increase in the use of such radioactive isotopes as tracers for the study of the behaviour of the stable elements, especially in physiology and diagnostic medicine. The increase in the number of those practicing nuclear medicine, including nuclear medicine technicians, has been correspondingly large. At least 10,000 hospitals or medical laboratories in the United States have facilities for the assay and use of radiopharmaceuticals.

It has been the purpose of this book to introduce, chronologically and historically, the concepts of radioactivity in a rather elementary way that will be of use to those who have received no extensive training in nuclear physics, but who nevertheless must make radioactivity measurements in the practice of nuclear medicine. Such concepts include the radioactive decay law, parent-daughter relationships such as that of ^{140}Ba–^{140}La, the interactions of radiation with matter by which these radiations are detected, assayed, and identified, the Einstein mass-energy relation and its bearing on the possible mode of decay (α, β^-, β^+ or electron capture), and the production of radioactive materials by nuclear bombardment. Some other matters, such as calorimetry not directly related to nuclear medicine, have also been included to make a consistent and, it is hoped, an interesting whole.

The closing chapters of this book will be concerned with the detection of radiation through its interaction with matter, the elementary principles of electronic equipment that has been developed to process and record radioactivity data, the statistics of counting, and the identification of radioactive isotopes of the elements by photon spectrometry. These are the fundamentals underlying most methods of measurement used in nuclear-medicine laboratories.

We wish to express our deep gratitude to a number of friends and colleagues for most helpful discussions, suggestions and comments on various parts of this book. They are: Dr. John H. Hubbell for reading and commenting on part of Chapter 3; Professor Ugo Fano for his comments on Chapter 5 and part of Chapter 6; Dr. Dale D. Hoppes for discussions of many parts of Chapters 6 and 8; Mr. John G. V. Taylor and Mr. Robert J. Toone for their comments on the section of Chapter 6 dealing with solid-state detectors; Dr. Keith Eberhardt and Dr. Harry H. Ku for their discussion of the section on statistics in Chapter 8; and Dr. Martin J. Berger and Dr. J. Joseph Coyne for several discussions concerning the Fano factor.

Only a few of the landmark references are given in this book to the early work in radioactivity. For a more comprehensive bibliography, of some two hundred references, the reader may wish to consult a very interesting Mound Laboratory Report (MLM-1960, September 22, 1972) entitled "The Early History of Radiochemistry" by Dr H. W. Kirby. This report describes the procedures first used to isolate many of the elements in the radioactive series.

W. B. MANN
R. L. AYRES

April 1979

1 The Discovery of Radioactivity and Early Experiments Into Its Nature

RADIOACTIVITY, the name coined by Marie Curie to describe the outward manifestations of atomic transformation, is the subject of this book.

Three discoveries in the closing years of the nineteenth century were not only to have a profound effect on scientific progress and thought in the first half of the twentieth century, but were also to create a new pattern for international political power and even, possibly, for life itself. These discoveries were those of x rays by W. C. Röntgen in 1895, of radioactivity by A. H. Becquerel in 1896, and the discovery, or postulation, of the corpuscular nature of the cathode rays by J. J. Thomson in 1897. In 1900 Max Planck also put forward the fundamental concept of the quantum in order to meet the failure of classical theory to describe the spectral distribution of black-body radiation; the so-called "ultraviolet catastrophe." Thus in the brief space of five exciting years were laid some very important foundations for the nuclear era upon whose threshold the world now stands!

The discovery of radioactivity was a direct consequence of that of the x rays or Röntgen rays. These had been produced in a discharge tube by the action of the cathode rays impinging upon the glass walls of the tube itself. A strong luminescence of the glass thus accompanied the production of the x rays, and there was considerable speculation whether such luminescence, or fluorescence, might of necessity be associated with the production of x rays. This possibility thus led to an intensive search for x-ray emission by materials rendered fluorescent or phosphorescent by visible light.

In the course of this search Antoine Henri Becquerel experimented with a double sulfate of uranium and potassium, a substance whose

1

phosphorescence had been investigated previously by him and his father, Alexandre Edmond Becquerel. In the session of February 24, 1896, of the Académie des Sciences of Paris, Becquerel presented a short note announcing that he had carried out experiments using a crystal of this double sulfate of uranium and potassium, in the form of a thin transparent "crust." In these experiments he took a photographic plate that was wrapped in black paper of such thickness that a day's exposure to the sun did not fog the plate. He then placed the fluorescent salt on the outside of the black paper and exposed the whole package to sunlight for several hours. On development, the silhouette of the fluorescent crystal appeared in black upon the photographic plate. Further, on interposing a piece of money or a metal plate pierced with a design, its outline also appeared on the plate. Becquerel also interposed a thin sheet of glass between the fluorescent salt and the paper wrapping of the plate to exclude the possibility of chemical action from the vapors that might result from heating in the solar radiation. (In retrospect we now know that the effect observed by Becquerel must have been caused chiefly by the β rays, as the less penetrating α particles could not have passed through the black paper in which the plate was wrapped.)

At the session of the Académie one week later, on Monday, March 2, 1896, Becquerel reported a most significant development. On the preceding Wednesday and Thursday he had prepared experiments similar to those which he had described at the beginning of that week. But, and so typical of that time of the year, the sun only appeared intermittently, and so he had returned the wrapped photographic plates to the darkness of a drawer, leaving in place the small crystalline layers of uranium salt. As the sun failed to appear on the following days, he developed the plates on March 1, expecting to find quite weak images. "The silhouettes appeared, to the contrary, with great intensity." Becquerel had discovered that the effect on the photographic plate was independent of the phenomenon of fluorescence which, he pointed out, could scarcely be perceived after one-hundredth of a second. In this communication he also referred to the "*radiations actives*," thereby anticipating the term "radioactivity" used later by Mme. Curie.

Subsequently, Becquerel reported that all compounds of uranium that he had examined, whether fluorescent or not, and uranium itself, exhibited this phenomenon. He also showed that metallic uranium was more active than its compounds, when disposed over roughly the same surface area, and that the radiations were weakened by interposing thin foils of aluminum, copper, and other absorbers between the source of radiation and the photographic plate, as witness the earlier experiments giving the silhouettes of money and other objects. Becquerel also demonstrated that these radiations, like the x rays, possessed the property of discharging electrified bodies; it is this property that was and is the basis of most methods for detecting radiation from radioactive substances. Becquerel also thought that he had been able to demonstrate the reflection, refraction, and polarization through tourmaline of these radiations, but these results could not be repeated by other physicists at that time. It is possible that some of Becquerel's results, particularly that of the reflection experiment, could have been accounted for by the back-scattering of radiation. Finally, Becquerel kept a piece of uranium for several years in darkness at the end of which time its action on a photographic plate was essentially unaltered. He thus demonstrated that these radiations from uranium occurred independently of the phenomenon of luminescence.

Apart from the discovery of radioactivity itself, one of the most important contributions made by Becquerel was to show that the radiations from uranium would cause an electrified body in air to be discharged, and he used an elementary form of electroscope, designed by Dragomir Hurmuzesen, to make semiquantitative measurements of the radioactive intensity. This method is considerably faster and more convenient than that of observing the photographic effect.

Two years after the discovery of radioactivity, Marie Sklodowska Curie turned her attention to the possibility of there being other radioactive elements. She therefore examined a large number of materials for radioactivity, using for this purpose the electric method of measurement in conjunction with the phenomenon of piezoelectricity which had been discovered by the brothers Jacques and Pierre Curie, the latter of whom she had married in 1895. Her apparatus consisted essentially of a parallel-plate condenser (AB in Fig. 1-1) on the lower plate of which the material under test was placed in the form of a layer of powder, the thickness of this layer being from 0.25 to 6.0 mm. The diameter of each of the plates was 8.0 cm, and they were usually 3 cm apart. The lower plate B of this condenser was connected to one pole of a

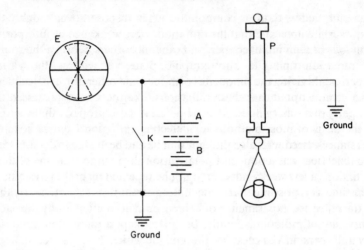

FIG. 1-1   Schematic diagrams of apparatus used by Pierre and Marie Curie to test the radioactivity of materials. The material to be tested was placed on the lower plate B of the parallel-plate condenser AB. The deflection of the electrometer E was maintained at zero by loading the quartz piezoelectric crystal P at a suitable rate with weights. K is a grounding key which is opened at zero time.

source of potential, the other pole of which was connected to ground. The other plate, A, of the condenser was connected to one pair of plates of a quadrant electrometer and also to one electrode of a piezoelectric crystal, of quartz, the other electrode of which was grounded. Under the influence of the radiations from radioactive materials the air between the plates A and B became ionized and conducting, and a current, of the order of 10^{-11} ampere, flowed between the plates. One pair of quadrants the electrometer E thus acquired an electric potential which would, in turn, cause the electrometer needle to be deflected. On now applying weights to the piezoelectric crystal a compensating potential is developed which neutralizes that due to the current across the condenser, and the electrometer deflection can be reduced to zero. In this condition the mass added in a given time gives a measure of the current flowing across the condenser as a result of the air being rendered conducting by the action of radiation. At the beginning of any measurement, plate A and the pair of quadrants to which it is connected are grounded by means of the key K. On increasing the voltage on plate B the current reaches a limiting value, called the *saturation current*, which is a rough relative measure of the radioactive intensity of the layer of material on plate B.

Using this electrical method, Marie Curie examined a great many chemical compounds and also a large number of rocks and minerals to see whether any other elements exhibited the phenomenon of radioactivity found in uranium. As a result she found that thorium was the only element which was radioactive to the same degree as uranium, a result which was obtained independently by G. C. N. Schmidt and published a few weeks earlier by him. Marie Curie also showed that the radioactivity of compounds containing uranium and thorium was in proportion to the amounts of these elements present. As it was also found to be independent of any change in physical state or chemical decomposition, she concluded that radioactivity was an atomic phenomenon.

Mme. Curie also found, however, that some minerals, notably pitchblende, chalcolite, autunite, and carnotite, were more radioactive than uranium itself. Thus measurements with these materials gave the following currents, in 10^{-11} ampere, across the condenser AB: uranium, 2.3; pitchblende from various localities, 1.6 to 8.3; chalcolite, 5.2; autunite, 2.7; carnotite, 6.2.

The question then arose: If radioactivity were an atomic phenomenon, how could minerals containing uranium, and perhaps also some traces of thorium, be more radioactive than pure uranium itself? To test this point Mme. Curie prepared crystals of artificial chalcolite, which is a double phosphate of copper and uranium, also known as tobernite. This artificial chalcolite exhibited a normal radioactivity given by its composition, namely some two-and-a-half times less than that of uranium. She concluded therefore that these minerals must contain traces of some radioactive element many times more active than uranium itself.

Pierre and Marie Curie therefore set themselves the task of isolating this hypothetical element which had no known property except its radioactivity. Their procedures could therefore be based only on the methods of chemistry supplemented by observation of the radioactivity, and their investigation was, in a sense, the first application of the techniques of radiochemistry. Thus they observed the radioactivity of a compound which was then subjected to chemical decomposition, and the radioactivities of all the products were determined. In this way the radioactive substances could be followed through the various stages of the chemical process.

Eventually two new radioactive elements were discovered, the one chemically similar to bismuth and separating with it, and the other chemically similar to, and separating out with, barium. The former they named polonium (in honor of Poland, the country of Mme. Curie's birth) and the latter, which they discovered in collaboration with G. Bemont, radium. In their paper in 1898 describing the discovery of radium, they comment that both polonium and radium when placed near a luminescent salt, render it fluorescent. They go on to comment that this weak source of light which functions without a source of energy (as they then thought) is in contradiction to the Carnot Principle.

It was now important to be sure that these activities that separated with bismuth and radium should be identified as being due to new elements. This proved difficult in the case of polonium, but by repeated fractional crystallization, radium chloride, which is less soluble than barium chloride, was eventually concentrated to a point where E. Demarçay was able to identify spectroscopically first one and then more new emission lines due to the new element. The first new line identified was that of greatest intensity at 3814.8 Å (381.48 nm) in the ultraviolet.

Finally, in collaboration with A. Debierne and with the help of a gift of one ton of residues from Joachimstal pitchblende by the Austrian Government, Mme. Curie was able to separate about 8 kg of barium and radium chloride having an activity of some sixty times that of metallic uranium. Finally, working with only about 0.1 g of radium chloride which had been found by Demarçay spectroscopically to have only the slightest trace of barium, Mme. Curie determined the atomic weight of radium and found it to be 225. The presently accepted value of this quantity is 226.05, as determined by O. Hönigschmid in 1934. In 1903 the Nobel Prize in Physics was awarded to Pierre and Marie Curie and Henri Becquerel for their work. In 1911 the Nobel Prize in Chemistry was also awarded to Marie Curie for isolating radium. She died in 1934 as a result of prolonged exposure to the radiations from the element which she had successfully isolated.

Eventually many more naturally occurring radioactive elements were discovered and their relationships to each other largely established by the work of Ernest Rutherford and Frederick Soddy. To Becquerel and the Curies belong the credit of the discovery of radioactivity and the elements polonium and radium, but in the next four decades the field was dominated by Rutherford and his colleagues. By 1904 some 20 radioactive elements had been discovered, mainly by the work of Rutherford in a series of brilliant and incisive researches, often in collaboration with Soddy. It was also a happy accident in this connection, as will be explained later, that P. Curie discovered an "emanation" from radium and Rutherford and Soddy an "emanation" from thorium and that the latter investigators were able to conclude that these emanations were rare gases, namely, radon and thoron (recognized later as an isotope of radon). It was also found that there exist three "families" of radioactive elements.* The parent of the third family, actinium, was discovered by Debierne in 1899 and independently by F. Giesel in 1901. Rutherford showed, by investigating their penetrating power, that there were two types of radiation emitted by uranium. These were the α rays, which could easily be stopped by a sheet of paper or a few centimeters path in air (the "range" in air), and the β rays, which were far more penetrating and could pass through several millimeters of aluminum. In 1900 a third, much more penetrating type of radiation was found, by P. Villard, to be emitted by radium, and the rays comprising this type were named γ rays.

As early as 1899 F. Giesel, S. Meyer and E. von Schweidler, and Becquerel showed that the β rays could be deflected in a magnetic field in the same sense as the cathode rays. By a deflection through 180° onto a photographic plate (P in Fig. 1-2), Becquerel also showed that the spectrum of the β rays varies continuously in energy, and he also demonstrated that those which impinge upon the plate nearest to the source, S, are most readily absorbed. This he showed by placing different absorbers of paper, glass, and metals directly on the photographic plate so that different amounts of the continuous spectrum were suppressed. Becquerel and E. Dorn showed in 1900 that the β rays could also be deflected, in the same manner as cathode rays, by an electric field. In the

* These are described in more detail in Chapter 2 and illustrated in Figs. 2-3, 2-4, and 2-5.

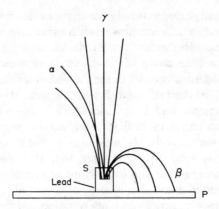

FIG. 1-2   Schematic of the paths of α, β, and γ radiation under the influence of a magnetic field at right angles to the plane of the drawing. The radioactive source S is contained in a lead cup. By allowing the β rays to fall on the photographic plate P, Becquerel showed that their energies varied continuously from zero, for any given substance, to a maximum.

same year Pierre and Marie Curie confirmed, by collecting them in a Faraday cylinder, that the β rays are negatively charged, while Becquerel, by observing their deflections in magnetic and electric fields, showed that β rays, having a velocity of some 1.6×10^{10} cm s^{-1}, have an approximate ratio of charge to mass ($e/m = 10^9$ C kg^{-1}) of the same order of magnitude as that for the cathode rays. Also in 1900, W. Kaufmann showed that for β rays having velocities between 2.36×10^{10} and 2.83×10^{10} cm s^{-1}, the ratio of charge to mass varies from 1.31×10^9 to 0.63×10^9 C kg^{-1}, respectively, thereby also indirectly demonstrating the relativistic dependence of mass on velocity. As the corresponding values for cathode rays, determined by J. J. Thomson in 1897, were 3×10^9 cm s^{-1} and 0.7×10^9 C kg^{-1}, it was readily concluded that the β rays were of the same nature as the cathode rays, which were negatively charged and, according to Thomson, corpuscular in form.

  The identity of the α and γ rays eluded elucidation for a great many more years. Initially, they were both believed to be unaffected by an electromagnetic field, but in 1902 Rutherford and A. G. Grier showed that the α rays were indeed deflected in powerful electric or magnetic fields. For a given field their deflection was much less than that for the β rays and in an opposite direction, indicating that the α rays consist of positively charged particles. Also in 1902, by measuring their deflection in electric and magnetic fields, Rutherford concluded that the α rays from radium were travelling with a velocity of about $2.5 \times 10^9 \,\mathrm{cm\,s^{-1}}$ and that they had a charge-to-mass ratio of about $6 \times 10^5 \,\mathrm{C\,kg^{-1}}$. The method used by Rutherford to make these measurements is illustrated in Fig. 1-3. In 1903, T. des Coudres, using the less sensitive photographic method of detection and radium bromide as a source, obtained values of velocity equal to $1.65 \times 10^9 \,\mathrm{cm\,s^{-1}}$ and e/m equal to $6.4 \times 10^5 \,\mathrm{C\,kg^{-1}}$ for the α rays from radium. Unlike the β rays, which were found to have continuously varying velocities (Fig. 1-2), the α particles from any given radioactive element were found to have discrete velocities and, hence, to be homogeneous in energy.

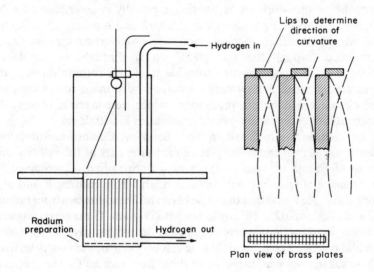

FIG. 1-3   Apparatus used by Rutherford for the magnetic and electrostatic deflection of the α rays from radium. Twenty to twenty-five parallel brass plates insulated from each other by ebonite were disposed below a gold-leaf electroscope in the manner shown. By applying a magnetic field normal to the plane of this page or an electric field between the plates, the α rays could be made to travel in arcs of circles. By increasing the field strengths until the α rays just failed to enter the electroscope, the respective limiting radii of curvature could be determined, and hence mv/e and mv^2/e could be calculated. The direction of curvature was determined by using plates with lips as illustrated. Radium emanation was flushed from the apparatus by a stream of hydrogen.

Helium had been discovered on earth in 1895 by W. Ramsay, its spectrum having first been observed in the solar spectrum by J. N. Lockyer. In 1903 Ramsay and Soddy showed conclusively that helium is produced from radium. Previously, Rutherford and Soddy had pointed out that one could speculate on whether the invariable presence of helium in minerals containing uranium and radium might be connected with their radioactivity. In the experiments of Ramsay and Soddy radium bromide in solution was used to supply radium emanation to a vacuum system containing a discharge tube to excite the helium spectrum, if helium were present. Initially, no helium spectrum could be observed after condensing the emanation in liquid air, but on repeating the experiment after the equipment had stood for four days the characteristic lines of the helium spectrum could be observed. The accounts of this work are fascinating, especially in describing how the liquid air trap shone brilliantly in the dark and how, on removing the liquid air, a brilliant phenomenon was observed as the emanation expanded through the equipment—slowly through the capillaries, delayed by the phosphorus pentoxide trap, and rushing through the wider glass tubes. A charming reference is also made to the vacation which intervened and the precautions which were taken to prevent the apparatus bursting while gases accumulated for 60 days!

In 1906, in a paper written from Berkeley, California, Rutherford describes further and more precise measurements of the velocity and ratio of charge to mass of the α rays from radium C, obtained by exposing a wire to radium emanation, and from radium F and also actinium.* He found that the value of e/m for the α particles from radium C was equal to 5.07×10^5 coulombs per kilogram. The experiments with radium F and actinium were less precise, but as they gave rather closely similar results he concluded that the values of e/m for the α particles from these substances were the same as those from radium C. The results of experiments carried out at the same time in collaboration with O. Hahn on the α particles of thorium showed that these also had a similar charge-to-mass ratio.

As it was natural to assume that all the particles of the α rays carried equal charges, it therefore seemed probable that all the α particles were

---

*These four naturally occurring α emitters are listed in Figs. 2-3 and 2-5.

identical in mass. Furthermore, the value of e/m for the hydrogen ion liberated in the electrolysis of water was $9.63 \times 10^5 \, C \, g^{-1}$ (i.e., the faraday).* If it were assumed that the charge on the α particle was equal to that carried by the hydrogen ion, then the mass of the α particle would be equal to about twice the mass of the hydrogen atom (or the same as that of the hydrogen molecule). If, however, the charge on the α particle were twice that of the hydrogen ion, then the mass of the α particle would be approximately equal to that of the helium atom.

In the light of the work of Ramsay and Soddy and his own observation with Soddy of the association of helium with uranium and radium minerals, Rutherford was inclined to accept the second view as to the identity of the α particle, namely, that it was a helium atom rather than a hydrogen molecule. Support for this point of view came when Rutherford and H. Geiger, in 1908, measured both the number of α particles emitted in a given time by radium C and their charge, and then related these quantities to the corresponding values for 1 gram of radium. In this investigation the first Geiger counter was used to measure the numbers of α particles emitted in a given time, and in their paper the authors discuss the statistics of counting. Rutherford and Geiger found that the average number of α particles emitted per second by 1 gram of radium was 3.4×10^{10} and that the charge on the α particle was 9.3×10^{-10} esu (i.e. 3.1×10^{-19} C).

The best values of the electronic charge, e, available at that time were those of J. J. Thomson, H. A. Wilson, and R. A. Millikan and L. Begeman and were respectively 1.03×10^{-19}, 1.35×10^{-19}, and 1.55×10^{-19} C. Rutherford and Geiger pointed out that the charge on the α particle, if it were assumed to be an integral multiple of e, must therefore be equal to $2e$ or $3e$. But they also gave reasons for believing that the then available values of e were too low and concluded, therefore, that the charge on the α particle is $2e$. From this deduction they proceeded to give a value of e, as determined by radioactivity measurements, which was equal to one-half their value of the charge on the α particle, namely, 1.55×10^{-19} C. As the presently accepted value for e is 1.602×10^{-19} C, which is equal to 4.80×10^{-10} esu, their value was surprisingly accurate. Proceeding from their deduction that the

* The current value of the "Faraday constant" is $9.648456 \times 10^4 \, C \, mol^{-1}$.

charge on the α particle is twice that on the hydrogen ion, Rutherford and Geiger then concluded (using the 1906 value of the charge-to-mass ratio of the α particle of 5.07 × 10³ emu per gram and taking the faraday to be 9.63 × 10³ emu per gram) that the atomic weight of the α particle was 3.84. Since the atomic weight of helium at that time was accepted as 3.96, the authors concluded that "taking into account probable experimental errors in the estimates of the value of e/m for the α particle, we may conclude that *an α particle is a helium atom*, or, to be more precise, *the α particle, after it has lost its positive charge, is a helium atom*."

The final and conclusive evidence in the quest for the identity of the α particle was provided in 1909 by the experiment of Rutherford and T. Royds in which radium emanation was confined in a thin-walled glass tube within a concentric outer vacuum discharge tube, the glass of the inner tube being only one-hundredth of a millimeter thick (which is equivalent to an α particle range in air of about 2 cm as compared with about 4 cm for radon and 4.6 cm for radium A). The vacuum tube was completely evacuated. Two days after introduction of the emanation into the thin-walled glass tube the yellow line of helium could be observed spectroscopically on exciting the discharge tube. After a further four days all the stronger lines of the helium spectrum could be observed. In control experiments helium was introduced into the inner thin-walled glass tube, but was found not to penetrate into the discharge tube. It was therefore concluded that the α particles from the decay of the emanation were the source of the helium observed in the outer discharge tube.

The nature of the γ rays eluded clarification for several more years. It had not been found possible to deviate them in magnetic or electric fields, and they were the most penetrating of the radioactive radiations. Becquerel in 1903 mentioned that these nondeviable rays had many common characteristics with the x rays. Rutherford and C. G. Barkla also inclined to the view that they were similar in nature to x rays, but others, notably F. Paschen and W. H. Bragg, felt that they were particles travelling at high velocities. As early as 1896 L. Fomm had shown that x rays could be diffracted by a slit, but attempts to repeat this experiment proved unsuccessful. It was not until 1912 that their electromagnetic nature, similar to light but of shorter wavelength, was completely established by the experiments of M. von Laue with W. Friedrich and P. Knipping and by those, in 1913, of W. H. Bragg and W. L. Bragg. In 1914

Rutherford and E. N. da C. Andrade carried out similar diffraction experiments with γ rays using a rock-salt crystal and showed, not only that the γ rays were similar in nature to the x rays, but they also measured the wavelengths of a number of γ rays emitted by radium B and radium C.

It is interesting in this connection to speculate what might have been the course of events had the electron been discovered first as a wave rather than a particle, or if the experiments of C. J. Davisson and L. H. Germer and of G. P. Thomson had preceded those of J. J. Thomson.

The literature of the era in those aspects of radioactivity that have been covered by this chapter makes exciting and comprehensible reading.

References

Becquerel, A. H. (1896) Sur les radiations emises par phosphorescence, *Comptes Rendus, Académie des Sciences, 122*, 420.

Becquerel, A. H. (1896) Sur les radiations invisibles émises par les corps phosphorescents, *Comptes Rendus, Académie des Sciences, 122*, 501.

Becquerel, A. H. (1896) Sur les radiations invisibles émises par les sels d'uranium, *Comptes Rendus, Académie des Sciences, 122*, 689.

Becquerel, A. H. (1896) Sur les propriétés different des radiations invisibles émises par les sels d'uranium, et du rayonnement de la paroi anticathodique d'un tube Crookes, *Comptes Rendus, Académie des Sciences, 122*, 762.

Becquerel, A. H. (1896) Emission de radiations nouvelles par l'uranium métallique, *Comptes Rendus, Académie des Sciences, 122*, 1086.

Curie, Sklodowska (1898) Rayons émis par les composés de l'uranium et du thorium, *Comptes Rendus, Académie des Sciences, 126*, 1101.

Curie, P. and Curie, S. (1898) Sur une substance nouvelle radio-active contenue dans la pechblende, *Comptes Rendus, Académie des Sciences, 127*, 175.

Curie, P., Curie, M. and Bémont, G. (1898) Sur une nouvelle substance fortement radioactive, contenue dans la pechblende, *Comptes Rendus, Académie des Sciences, 127*, 1215.

Debierne, A. (1899) Sur une nouvelle matière radio-active, *Comptes Rendus, Académie des Sciences, 129*, 593.

Debierne, A. (1900) Sur un nouvel element radio-actif: l'actinium, *Comptes Rendus, Academie des Sciences, 130*, 906.

Demarçay, E. (1898) Sur le spectre d'une substance radio-active, *Comptes Rendus, Académie des Sciences, 127*, 1218.

Geiger, H. and Rutherford, E. (1912) Photographic registration of α particles, *Phil. Mag., 24* [6], 618.

Giesel, F. (1899) Über die Ablenkbarkeit der Becquerestrahlen in magnetische Felde, *Ann. d. Phys., 69* [3], 834.

Meyer, S. and v. Schweidler, E. R. (1899) Über das Verhalten von Radium und Polonium im magnetischen Felde, *Phys. Z., 1*, 90 and 113.

Ramsay, W. and Soddy, F. (1903) Experiments in radioactivity and the production of helium from radium, *Proc. R. Soc.*, A, *72*, 204.

Röntgen, W. C. (1895) Über eine neue Art von Strahlen, *Sitsungsberichte der Physikalisch-Medicinischen Gesellschaft zu Würzburg*, 132.

Rutherford, E. (1903) The magnetic and electric deviation of the easily absorbed rays from radium, *Phil. Mag.*, 5 [6], 177.

Rutherford, E. (1924) Early days in radioactivity, *Journal of the Franklin Institute, 198*, 281.

Rutherford, E. and Geiger, H. (1908) An electrical method of counting the number of α-particles from radio-active substances, *Proceedings of the Royal Society*, A, *81*, 141.

Rutherford, E. and Grier, A. G. (1902) Deviable rays of radioactive substances, *Phil. Mag.*, *4* [6], 315.

Rutherford, E. and Royds, T. (1909) The nature of the α particle from radioactive substances, *Phil. Mag.*, *17* [6], 281.

Schmidt, G. C. N. (1898) Über die von Thorium und den Thoriumverbindungen ausgehende Strahlung, *Verhandlungen der Deutschen Physikalischen Gesellschaft, 17*, 14.

2 Radioactive Change
and the Theory of
Successive Radioactive
Transformations

As has already been pointed out, it is a happy chance that a member of each of the three naturally-occurring radioactive families happens to be gaseous and, in addition, short-lived. By being gaseous, these radioactive products could effect their own chemical separation. By removing these gaseous radioactive daughters, Rutherford was soon able to show that a fresh supply of these daughters was always being generated by the decay of their long-lived radioactive parents (see Fig. 2-6). They could therefore scarcely be missed!

The effects of such radioactive gases produced by thorium and by radium were noticed by R. B. Owens and by P. and M. Curie in October and November 1899, respectively. In Owens' experiments the radioactive compound was placed on the lower plate of a parallel-plate condenser, and the ionization current, which flowed when the lower plate was raised to a positive potential of 95 volts, was measured by means of an electrometer connected to the upper plate. The parallel-plate condenser was mounted within a metal box. On passing a current of air through this box, the ionization between the condenser plates dropped, in the case of thorium oxide, to 33 per cent of the ionization current in still air. The Curies observed that preparations containing radium were able to transfer radioactivity to inactive substances in their vicinity or in the same enclosure. Shortly thereafter, in 1900, Debierne observed the same phenomenon of transferred radioactivity with preparations containing actinium.

The Curies also observed that this transferred activity, which they called *induced* radioactivity, not only increased with the time of exposure

15

to the radium preparation, but that it also decayed, very rapidly at first, on removing the activated material from proximity to the radium. These induced radioactivities were as much as 50 times that of uranium. While they continued to investigate this phenomenon, the Curies never fully understood how it arose.

Rutherford quickly realized the importance of the dependence of the apparent radioactivity of thorium on the movement of air. Owens' results appeared in the *Philosophical Magazine* of October 1899. In the January 1900 issue of the same journal Rutherford published a paper on "a radioactive substance emitted from thorium compounds" and one month later another paper in which he described experiments in which the "excited" activity could be concentrated onto wires which had a negative potential relative to their surroundings. This excited radioactivity, which Rutherford states was first noticed because the insulators (such as ebonite and paraffin) lost their insulating properties in the presence of thorium compounds, he concluded to be directly caused by the "emanation" from the thorium.

One of the experiments described by Rutherford in his January 1900 paper is so artistically simple, and so beautifully illustrative of what the scientific method should be, that it is well worth describing briefly. His equipment is illustrated in Fig. 2-1. A current of air passes over a thick layer of thorium oxide contained in the paper container *A* in the long

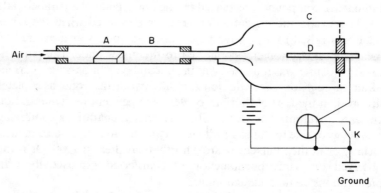

FIG. 2-1   Rutherford's apparatus to demonstrate the decay of thoron emanation. Thorium oxide is contained in a paper box at *A*.

metal tube B. Thence the air passes into a large cylindrical insulated vessel C from which it flows through a number of small holes. An insulated metal electrode D is connected to one pair of quadrants of a Kelvin electrometer, as shown. After the air had been allowed to flow for some time it was stopped, and the ionization current between C and D was measured for some ten minutes. The results obtained are illustrated by curve A in Fig. 2-2. The beauty of the experiment, as also the insight of

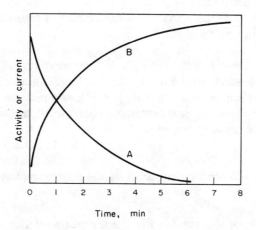

FIG. 2-2   Decay and growth curves (A and B respectively) of thoron.

the investigator, lies in its second part. Rutherford now placed the thorium oxide, in its paper container, between two concentric insulated metal cylinders along which a current of air was passed in order to remove the emanation as soon as it was formed.* On stopping the flow of air and on measuring the ionization current between the two cylinders, the results represented in curve B of Fig. 2-2 were obtained. Curves A and B have been normalized so that the maximum of B is equal to the initial value of A and their sum at any time t is seen to be constant. This

* This experiment could have been carried out with the paper boat in the ionization chamber C in Fig. 2-2, but Rutherford does not make it clear in his paper whether he used this or another apparatus.

indicated that at equilibrium the emanation was being produced as fast as it disappeared, and Rutherford concluded that if n were the number of ions produced per second at time t by the radioactive disintegrations occurring in the air contained between the cylinders and q were the number of ions supplied per second by the emanation diffusing from the thorium, then the rate of variation of ions at time t was given by

$$dn/dt = q - \lambda n,$$

where λ is a constant.

From this he deduced that if i is the current at time t and i_m is the maximum current, then

$$i/i_\mathrm{m} = 1 - e^{-\lambda t}, \tag{2.1}$$

which is the form of curve B in Fig. 2-2. For large values of t, $i \to i_\mathrm{m}$, and for $t = 0$, the time at which the air flow was stopped, $i = 0$.

If, however, the source of the emanation be removed, then $q = 0$ and the rate of ion production becomes

$$dn/dt = -\lambda n.$$

If $i = i_0$ at $t = 0$, the time of removing the source, we have

$$i/i_0 = e^{-\lambda t}, \tag{2-2}$$

which is the form of curve A in Fig. 2-2, the measured ionization current between C and D in Fig. 2-1 decreasing in geometrical progression as a function of time. If we normalize Eqs. 2-1 and 2-2 so that $i_\mathrm{m} = i_0$, then the total value of i for curves A and B at any time t is $i_0(1 - e^{-\lambda t}) + i_0 e^{-\lambda t} = i_0$. This is merely another way of saying that for equilibrium between a long-lived parent and a short-lived daughter (or short-lived daughter and granddaughters), the progeny are produced as fast as they decay.

In this paper Rutherford also described how the emanation behaved more like a gas in that it could be passed through plugs of cotton-wool and through hot or cold water and through strong or weak sulphuric acid without losing its radioactivity. He also showed in the paper of February 1900 that the rate of decay of the "excited activity" (which was that observed on solid bodies in the vicinity of the thorium) was less than that of the emanation, decreasing to about half its activity in 11 hours.

Moreover, the excited activity was due to a deposit which could be removed from a platinum wire by sulphuric acid, but was then found in the residue left on evaporating the acid to dryness.

With this introduction to give some idea of the scientific methods employed we can perhaps pass on more rapidly to the conclusion expressed by Rutherford and Soddy in a paper on "The Cause and Nature of Radioactivity," appearing in the November 1902 issue of the *Philosophical Magazine*, that "the interpretation of the above experiments must therefore be that the emanation is a chemically inert gas analogous in nature to the members of the argon family." In the same paper they speculate on the association of helium with radioactivity, and in its conclusion they state that "the law of the decay of activity with time ... then appears as the expression of the simple law of chemical change, in which one substance only alters at a rate proportional to the amount remaining."

From this "breakthrough" it was merely a matter of clearing up details so that, in 1908, when Rutherford and Geiger concluded that the α particle was identical with the helium atom, they were able to deduce, in the case of radium and using Mme. Curie's latest atomic weight of 226* for radium, that the atomic weight of radium emanation (radon) must be 222 and that of radium A (the active deposit form radon) must be 218.

The relationships within the three families of naturally-occurring radioactive substances, as they are now understood, are illustrated in Figs. 2-3, 2-4, and 2-5, where the differences between their atomic weight and atomic number are plotted against their atomic number. (The atomic number, a concept which is discussed later, is one, however, which will be familiar to most readers as the number of protons associated with the atomic nucleus. In 1913 H. G. J. Moseley, in his celebrated investigation of the wavelengths of the characteristic x rays emitted from the lighter elements, first used the term "atomic *number* of an element" to denote the numerical position of that element in the periodic table. He found that the frequencies of each of the series of characteristic x radiations were related approximately to the square of the atomic number. A. van den Broek, in January 1913, suggested that the nuclear charge of an element was proportional to the ordinal number which represented its position

---

* When given to the nearest whole number, this is also now called the *atomic mass number* or *nucleon number*.

in the periodic system, and this assumption was adopted shortly thereafter by both N. Bohr and Moseley.)

A further brief historical note is in order. It will be noticed in Figs. 2-3, 2-4, and 2-5 that many radioactive elements are identical chemically, but have different atomic weights. By 1911 some 40 radioactive

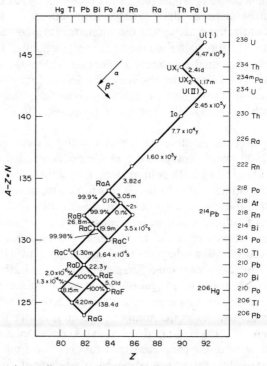

FIG. 2-3   Uranium-radium $(4n + 2)$ series. Half lives are shown in years (y), days (d), hours (h), minutes (m), and seconds (s). Values of $n = 59, 58, 57, \ldots 51$ give the atomic mass-numbers of this series. ^{234}Th decays to ^{234m}Pa (1.17m), 99.87 per cent of which decays to ^{234}U and 0.13 per cent to ^{234}Pa which decays to ^{234}U. The symbols for the early names of the radioactive elements are shown in the body of the figure. The artificially produced radioactive elements ^{246}Cm $(\alpha, 4730y)$ and ^{242}Pu $(\alpha, 3.87 \times 10^5 y)$ now also feed ^{238}U, and ^{242m}Am $(\alpha, \gamma, 152y)$ feeds both ^{238}U and ^{234}U through three different routes and a number of intermediate radioactive elements, namely ^{242}Am $(\beta^+, \beta^-, 16.02h)$, ^{238}Np $(\beta^-, 2.117d)$, ^{242}Pu, ^{242}Cm $(\alpha, 163.2d)$, and ^{238}Pu $(\alpha, 87.75y)$. For details see D. C. Kocher, 1977.

species were crowding a dozen positions that seemed appropriate to them in the periodic table. A number of investigators had also shown that several different radioactive substances were chemically inseparable. In the same year Rutherford proposed the concept of the nuclear atom, and Soddy, also in 1911, suggested that the emission of an α particle from a radioactive element created a new element in the periodic table in the next group or column *but one* in the direction of diminishing group number. Early in 1913 K. Fajans and Soddy independently concluded that the emission of an α particle resulted in an element two

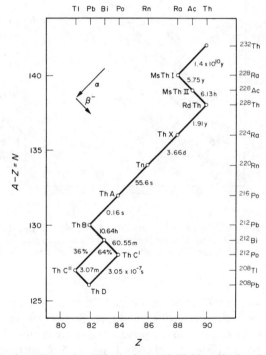

FIG. 2-4 Thorium (4n) series. Values of n = 58,57,56,...52 give the atomic-mass numbers of this series. The symbols for the early names of the radioactive elements are shown in the body of the figure. ^{220}Rn is still often referred to as thoron. ^{232}Th is now also fed by two series of artificially produced radioactive elements beginning with ^{252}Cf (α, 2.638y) and ^{244}Cm (α, 18.11y). ^{228}Th is also fed by ^{236}Pu (α, 2.851y) and ^{232}U (α, 71.7y). See D. C. Kocher, 1977.

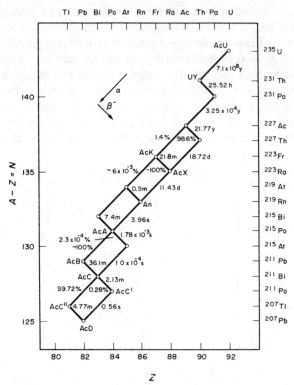

FIG. 2-5   Actinium (4n + 3) series. Values of n = 58, 57, 56, ... 51 give the atomic-mass numbers of this series. The symbols for the early names of the radioactive elements are shown in the body of the figure. ^{219}Rn is still sometimes referred to as actinon. ^{235}U is now also fed by a series of artificially produced radioactive elements beginning with ^{247}Cm ($\alpha, 1.56 \times 10^7$y). See D. C. Kocher, 1977.

columns "down" in the periodic table and the emission of a β particle in an element one column "up." In addition, Soddy concluded that all elements in the same place in the periodic table are chemically identical and cannot be separated by chemical means. For such elements he proposed the name "isotopes," meaning "in the same place."* Thus in the

---

* This term was apparently suggested by a physician, Dr. Margaret Todd while dining with Soddy (see "Frederick Soddy", by Alexander Fleck, *Biographical Memoirs of Fellows of the Royal Society*, 3, 202, 1957).

radioactive series illustrated in Figs. 2-3, 2-4, and 2-5 there are three isotopes of radium, having atomic weights 223, 224, and 226. There is also an isotope of radon in each of the families, and these have atomic weights of 219 (actinium emanation, or actinon), 220 (thorium emanation, or thoron), and 222 (radium emanation, or radon). These are now designated as radon-219 (^{219}Rn), radon-220 (^{220}Rn), and radon-222 (^{222}Rn), respectively. There are now some 27 known radioactive isotopes of radon ranging in atomic weight from 200 to 226. There are, however, as was later shown by the work of J. J. Thomson and F. W. Aston, many stable isotopes too. Thus we now know that hydrogen-2 (deuterium) is a stable isotope of hydrogen-1 while hydrogen-3 (tritium) is an isotope of hydrogen which is radioactive. There are ten stable isotopes of tin and at least eighteen radioactive isotopes of that element, ranging from 106 through 133 in atomic weight.

Nowadays the term "isotope" is often misused. Its meaning of "in the same place" signifies that it shares the *same position* in the periodic table. Thus phosphorus-30 is an isotope *of* phosphorus, and phosphorus-32 is a radioactive isotope, or radioisotope, *of* phosphorus. One should not refer to any atomic species as an isotope and leave out the word "of" after it. An atomic species is called a *nuclide* and a radioactive nuclide a *radionuclide*. Thus carbon-12 is a nuclide and carbon-14 is a radionuclide, and both are *isotopes of* carbon. Carbon-14 can also properly be termed a *radioisotope of* carbon.

In 1903, in a paper entitled "Radioactive Change" published in the *Philosophical Magazine*, Rutherford and Soddy propounded "the law of radioactive change." They pointed out that, in all cases that had been investigated, the activity of a radioactive material decreased in geometrical progression with time. Thus if the activity of a given substance decreased to half in a certain time, then, after a further equal interval of time, the activity would decrease by half again, or to one quarter the activity at the beginning of the first of the two equal intervals of time. They then deduced that the appropriate form of the law expressing such a geometrical decrease is that $N_t/N_0 = e^{-\lambda t}$, where N_0 is the number of "changing systems" (radioactive atoms) initially present, N_t is the number after time t, and λ is a constant. Differentiating, they obtained $dN_t/dt = -\lambda N_t$ and then stated that the "law of radioactive change may therefore be expressed in the one statement—The proportional amount of radioactive matter that changes in unit time is a constant." The constant λ they called the "radioactive constant," but it is now known also as the "decay constant" of a radioactive substance. The quantity 100λ $[= (100/N_t)(dN_t/dt)]$ is also clearly the percentage rate at which the number of radioactive atoms present at time t is decaying.

Rutherford and Soddy then observed that the law of radioactive change (the rate of change is proportional to the quantity of the changing substance) is analogous to the law of monomolecular chemical reaction. Nowadays it is usual to state that the law of radioactive decay is a statistical law, that the rate of change at time t $(-dN_t/dt)$ is proportional to the number (N_t) of radioactive atoms present. Thence by introduction of the constant of proportionality λ and integration with the boundary condition $N_t = N_0$ when $t = 0$, it is deduced that $N_t/N_0 = e^{-\lambda t}$. Either approach is a valid one; that of Rutherford and Soddy was based on experiment, while the statistical method was based on the work of E. von Schweidler in 1905.

This radioactive decay law applies to all radioactive substances, but the decay constant λ, which has the dimension T^{-1}, is different foɪ different radioactive elements. In addition to the three kinds of radiation that were observed at the turn of the century—namely, α, β, and γ—it has also since been found that radionuclides can decay by the emission of positrons (i.e., β^+ emission). In this event the daughter element is one down in atomic number or in the periodic table. As an alternative this same transition, resulting in the loss of one in atomic number, can be effected by the capture into the nucleus of an orbital electron with consequent emission of a characteristic x ray. This excitation energy can also interact with the outer electrons in an atom to give them sufficient kinetic energy to leave the atom as low-energy electrons, or Auger electrons as they are called after the discoverer of this effect. A further disintegration product arises from the process known as internal conversion. This process arises when, after the transition of a radioelement by the emission of a particle such as an alpha particle or of an electron or positron, or by orbital-electron capture, the daughter nucleus is left in an excited state, that is to say with an excess amount of energy which it then proceeds to get rid of in the form of a gamma ray. Internal conversion arises when instead of a gamma ray being emitted from the nucleus the excess energy is transferred to an orbital electron which leaves the atom with a velocity corresponding to the energy balances available in the transition.

This slight digression on the modes of decay is perhaps a little premature and a departure from the ordered chronological presentation. The details are discussed further at later stages in the development of our subject. The digression is, however, necessary in order to qualify an important statement which can be made about the decay constant λ of a radionuclide, namely, that it is a most fundamental characteristic of a radioactive substance, unaffected by practically all prevailing physical and chemical states of that substance and most useful in helping to identify radioactive elements. If, however, the decay involves the orbital atomic electrons, as in either orbital-electron capture or internal conversion, then, quite clearly, any changes which affect the proximity of these electrons to the nucleus will also affect the rate of decay. Thus in these two cases quite small changes can be effected by changes in chemical composition, and much larger changes in λ may occur in conditions of complete ionization which may exist in stars. Otherwise the decay constant is generally a unique characteristic of a radionuclide, as is also a quantity known as the half life of the radioactive substance, which is inversely proportional to λ. In the case of α and β emission, both λ and the half life are completely independent of the physical and chemical states of radioactive substances, as they exist on earth.

The half life, a concept introduced by Rutherford in 1904, is the time required for a radioactive element to decay from a given activity to half that activity. Since we have already noted that radioactivity decreases in a geometrical progression with time, it is clear that the time required for any given radioactive element to decrease in activity by a factor of $\frac{1}{2}$ is the same irrespective of the initial activity or the initial time.

Let us suppose, however, that the number of atoms of a radioactive substance present at time $t = 0$ is N_0. Then the time $t_{1/2}$ at which $N_t = \frac{1}{2}N_0$ is given by the radioactive decay law to be:

$$\tfrac{1}{2}N_0/N_0 = e^{-\lambda t_{1/2}}.$$

Thus the half life,
$$t_{1/2} = \frac{\ln 2}{\lambda}$$
$$= \frac{0.69315}{\lambda}.$$

In 1904 Soddy also introduced the concept of the mean life of a radioactive atom. Any given radioactive atom can disintegrate at any time between $t = 0$ and $t = \infty$. The number decaying in a time dt at t (as defined above) is $\lambda N\, dt$, which is equal to $\lambda N_0 e^{-\lambda t}\, dt$. The life of each of these atoms decaying at time t is *ipso facto*, t. Thus the average life of all the radioactive atoms is $\int_0^\infty t\lambda N_0 e^{-\lambda t}\, dt/N_0$, which is equal to $1/\lambda$. Because, as we have noted, radioactive decay follows a statistical law, the number of radioactive atoms decaying in a given time fluctuates within the normal statistical limits; λN_t, in fact, represents the *probability* that a number of atoms equal to dN_t will decay in the interval of time dt at t.

The half lives of radionuclides vary from less than 10^{-6} second to more than 10^{18} years, an enormously wide range.

An example of the use of the radioactive decay law can be given by calculating the number of atoms disintegrating per second in 1 gram of pure radium. The atomic weight of radium is 226.03 so that from the Avogadro constant, which is equal to 6.022×10^{23} per mole, the number of atoms N_0, at the time of separation, in 1 gram of radium-226 is $6.022 \times 10^{23}/226.03 = 2.664 \times 10^{21}$. The most recent value for the half life of radium-226 is 1600 years, so that the decay constant λ is equal to $0.69315/(1600 \times 3.156 \times 10^7) = 0.1373 \times 10^{-10}$ per second, 3.156×10^7 being the number of seconds in a year. The disintegration rate of the 1 gram of radium-226 is therefore given by $-dN_0/dt = \lambda N = 0.1373 \times 2.664 \times 10^{11} = 3.66 \times 10^{10}$ alpha particles per second. This quantity, for which early experiments gave values nearer to 3.7×10^{10} per second, was closely associated with the original definition of a unit of radioactivity, the *curie*, Ci, which was defined as the amount of radon in equilibrium with 1 gram of radium-226. This then became the basis of the later definition of the curie, as the activity of a radioactive substance such that it is decaying at a rate of 3.700×10^{10} disintegration, decays, or transitions, per second. The international system of units is based on the metre, the kilogram and the second as the units of length, mass and time, and is known as the *Système International* (SI). The present SI unit for the activity of a radionuclide is the *becquerel* (Bq), which is equal to one radioactive transition per second. The curie is, however, retained as a special SI unit for radioactivity.

In our treatment of the law of radioactive decay we have considered only the disintegrations of one radioactive element, or radionuclide. In the early days of radioactivity the investigators were, however, concerned with long series of radioactive transitions (see, for example, Figs. 2-3, 2-4, and 2-5). It was therefore of interest to know, not only how a parent radionuclide decays, but also the laws governing the amounts of succeeding generations of radioactive progeny present at any time. Rutherford derived the relationships between such members of a radioactive family and published them in 1904 in the *Philosophical Transactions of the Royal Society*.

One situation which was of particular interest to Rutherford was to expose a wire to radon for varying times and then determine the amounts of radium A, radium B, radium C, etc., on the wire as a function of time. His analysis therefore considered the number of atoms, $P, Q, R, S, \ldots$, of elements A, B, C, D, $\ldots$, present at time t under varying conditions of the supply of element A—e.g., element A alone was present at time $t = 0$, or element A was being supplied at a constant rate, and so on. He was thus able to derive the growth and decay curves for A, B, C, etc., in terms of their respective decay constants, λ_A, λ_B, λ_C, etc. He referred to this as the theory of successive radioactive transformations.

Nowadays we are not so interested in the naturally-occurring radioactive series of elements as in radionuclides which can now be artificially produced. Some of these artificially produced radionuclides decay to a radioactive daughter which in turn decays to a stable element. It is therefore more of interest at the present time to consider Rutherford's theory just for two related radionuclides, A and B.

Let us therefore suppose that the parent A only is present at time $t = 0$. We then desire to know the numbers P and Q of radioactive atoms of A and B, respectively, present at any subsequent time t. If P_0 is the number of atoms of A present initially, then at time t the radioactive decay law tells us that:

$$P = P_0 e^{-\lambda_A t}, \qquad (2\text{-}3)$$

and

$$dP/dt = -\lambda_A P, \qquad (2\text{-}4)$$

where λ_A is the radioactive decay constant for element A. Furthermore, if λ_B is the corresponding constant for element B, we know that at time t the number of atoms of B being created per second is $\lambda_A P$ (the number of A decaying to B) while the number of atoms of B disintegrating, in turn, is $\lambda_B Q$. At time t we can therefore state that

$$dQ/dt = \lambda_A P - \lambda_B Q \qquad (2\text{-}5)$$

(Similarly, if there had been further radioactive elements in the series,

$$dR/dt = \lambda_B Q - \lambda_C R, \text{ etc.})$$

If we substitute for P from Eq. 2-3 into Eq. 2-5, we obtain

$$dQ/dt = \lambda_A P_0 e^{-\lambda_A t} - \lambda_B Q.$$

To solve this differential equation we multiply both sides by the integrating factor $e^{\lambda_B t}$, whence

$$e^{\lambda_B t}(dQ/dt) + e^{\lambda_B t}\lambda_B Q = \lambda_A P_0 e^{(\lambda_B - \lambda_A)t},$$

from which we get

$$(d/dt)(Qe^{\lambda_B t}) = \lambda_A P_0 e^{(\lambda_B - \lambda_A)t}.$$

Integration of this equation with respect to t gives

$$Qe^{\lambda_B t} = \frac{\lambda_A}{\lambda_B - \lambda_A} P_0 e^{(\lambda_B - \lambda_A)t} + \text{constant}.$$

To obtain the constant of integration we can use the condition that $Q = 0$ at $t = 0$, so that the constant of integration is equal to $-\lambda_A P_0/(\lambda_B - \lambda_A)$. Thus

$$Qe^{\lambda_B t} = \frac{\lambda_A P_0}{\lambda_B - \lambda_A}(e^{(\lambda_B - \lambda_A)t} - 1),$$

or at time t,

$$Q = \frac{\lambda_A}{\lambda_B - \lambda_A} P_0(e^{-\lambda_A t} - e^{-\lambda_B t}). \qquad (2\text{-}6)$$

From Eq. 2-6 we note that $Q = 0$ for both $t = 0$ and $t = \infty$, which is to be expected for a radioactive daughter growing from a radioactive parent and then itself decaying with a half life of $(t_{1/2})_B$.

It is also apparent that, if Q is equal to zero for $t = 0$ and for $t = \infty$, then Q must also grow to a maximum at some intermediate time t_m. This time can be derived from Eq. 2-6 by determining the time at which $dQ/dt = 0$, that is, when

$$\lambda_A e^{-\lambda_A t_m} = \lambda_B e^{-\lambda_B t_m}$$

whence

$$t_m = \frac{\ln(\lambda_B/\lambda_A)}{\lambda_B - \lambda_A}. \tag{2-7}$$

When $dQ/dt = 0$, that is, when $t = t_m$, we have from Eq. 2-5 that

$$\lambda_A P = \lambda_B Q,$$

whence
$$dN_A/dt = dN_B/dt. \tag{2-8}$$

We see therefore that, whatever the relative half lives of the parent or daughter, their activities are *equal* at that time, t_m, that the activity of the daughter reaches a maximum value.

Great care must be taken not to confuse dQ/dt in Eq. 2-5 with the normal rate of decay of the daughter B. For times less than t_m, dQ/dt is positive, and for times greater than t_m it is negative. The instantaneous decay rate, or activity, of the daughter B at *any* time t is $dN_B/dt = -\lambda_B Q$. The decay rate of the parent A is, however, given by Eq. 2-4, and $dN_A/dt = -\lambda_A P = dP/dt$.

In the case which we have chosen to consider, that of a radioactive parent decaying to a radioactive daughter which in turn decays to a stable element, there are three variations of Eq. 2-6 which are of importance. Two of them are for cases where the parent is longer-lived than the daughter, and the third case is for when the daughter product is longer-lived than its radioactive parent. In the first instance one of the two cases occurs when the parent is very much longer-lived than the daughter so that $(t_{1/2})_A \gg (t_{1/2})_B$, or $\lambda_B \gg \lambda_A$.

Equation 2-6 can be rewritten as

$$\frac{\lambda_B - \lambda_A}{\lambda_A} Q = P_0 e^{-\lambda_A t} - P_0 e^{-\lambda_B t},$$

$$= P(1 - e^{-(\lambda_B - \lambda_A)t});$$

whence for $\lambda_B \gg \lambda_A$,

$$\lambda_B Q = \lambda_A P(1 - e^{-\lambda_B t}), \tag{2-9}$$

or

$$\frac{dN_B}{dt} = \frac{dN_A}{dt}(1 - e^{-\lambda_B t}). \tag{2-10}$$

Thus, if we separate a long-lived parent A from its daughter B by means, say, of a chemical separation, Eq. 2-10 shows how the activity of the daughter regrows with time. If we write the exponential in Eq. 2-10 as a function of the half life of the daughter, we have

$$\frac{dN_B}{dt} = \frac{dN_A}{dt}(1 - e^{-0.69315 t/(t_{1/2})_B}),$$

or, replacing $t/(t_{1/2})_B$ by $(n_{1/2})_B$, the *number* of half lives of the daughter B, then

$$\frac{dN_B}{dt} = \frac{dN_A}{dt}(1 - e^{-0.69315 (n_{1/2})_B}). \tag{2-11}$$

For $(n_{1/2})_B$ equal to one half life the exponential term in Eq. 2-11 equals 0.500, and for 2, 3, 4 and 5 half lives the exponential term is respectively 0.250, 0.125, 0.0625, 0.0313, and so on. Thus we see that after five half lives, the activity of the daughter has reached 97 per cent of that of the parent from which it had been separated. In a few more half lives the radioactivity of A is essentially equal to that of B because $e^{-\lambda_B t}$ in Eqs. 2-9 and 2-10 becomes negligible compared with unity, so that

$$\lambda_B Q = \lambda_A P = \lambda_A P_0 e^{-\lambda_A t}$$

and

$$\frac{dN_B}{dt} = \frac{dN_A}{dt}.$$

Thus for $\lambda_B \gg \lambda_A$, we see that the daughter B soon decays with the half life of the parent A and that their activities are equal. The ratio of the numbers of atoms Q to P is, however, in the inverse ratio of their decay constants λ_B and λ_A, but is proportional to their half lives. This is equivalent to saying that, for equal activities, it is necessary to have more atoms of the longer-lived radionuclide in proportion to its half life. In the same way, if different sources have numbers of atoms of different radionuclides in the same proportions as their half lives, they will all have equal activities.

The equilibrium represented by the above equations for $\lambda_B \gg \lambda_A$ is called *secular* equilibrium (Latin, *saeculum, an age*). Examples of such secular equilibrium are provided by the radioactive decay of radium-226 ($t_{1/2} = 1600$ years) to radon-222 ($t_{1/2} = 3.82$ days) and of strontium-90 ($t_{1/2} = 29.12$ years) to yttrium-90 ($t_{1/2} = 64.0$ hours). Both members of the former family emit α particles, and both of the latter emit negative β particles. The growth of radon-222 activity from freshly separated radium-226 is illustrated in Fig. 2-6. This also illustrates the growth of radon-222 activity in an enclosed solution of a radium-226 salt from which the radon has been removed by bubbling an inactive gas through the solution. From Eq. 2-7, it is also clear that the time t_m at which Q becomes a maximum becomes very long as λ_B/λ_A becomes very large.

An examination of Eq. 2-10 also explains the results obtained by Rutherford with thorium and thoron which are illustrated in Fig. 2-2. In

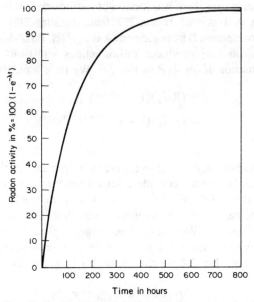

FIG. 2-6   Growth of radon-222 activity from freshly separated radium-226.

this case we are considering a gas, thoron (actually radon-220; $t_{1/2} = 55.6$ seconds), which is in equilibrium (through a series of solid daughter products) with thorium-232 ($t_{1/2} = 1.405 \times 10^{10}$ years). The sequence of radioactive products between the thorium-232 and the thoron (radon-220) is illustrated in Fig. 2-4. As secular equilibrium had long since been established, the thoron was being replenished as fast as it was being depleted by its own radioactive decay (i.e., $\lambda_A P$ for thorium-232, $\lambda_B Q$ for radium-228, $\lambda_C R$ for actinium-228, $\lambda_D S$ for thorium-228, $\lambda_E T$ for thorium-224, and $\lambda_F U$ for radon-220 are all equal). If one removed all, or part, of the thoron (radon-220), then the decay of the part removed must equal the rate at which new thoron is being produced. Thus the decay and growth curves, as shown in Fig. 2-2, exactly complement each other, and their sum is constant.

Another important application of Eq. 2-9 is in the production of radionuclides in a nuclear reactor, or by any other means such as nuclear or photonuclear bombardment, *provided that* the *rate of production R* of the radionuclide is *constant*. The process is then quite analogous to the practically constant rate of production of radionuclide B by a very long-lived parent A (such as radon-222 from radium-226). The rate of production of element B from element A is $\lambda_A P$ (cf. Eq. 2-5), so we can, in the case of artificially produced radionuclides, substitute the *constant* rate of production R for $\lambda_A P$ in Eq. 2-9. We then have that

$$Q = (R/\lambda_B)(1 - e^{-\lambda_B t})$$
$$= (R/\lambda_B)(1 - e^{-0.69315(n_{1/2})_B}), \qquad (2\text{-}12)$$

where $(n_{1/2})_B$ is the number of half lives for which the bombardment is continued. As before, five half lives give us 97 per cent of the maximum possible yield. Economic considerations often force one to settle for a bombardment of one half life (with a 50-per-cent yield), or even less!

The second case we were considering was that in which $\lambda_B > \lambda_A$, *but not very much greater*. We can then no longer neglect λ_A compared with λ_B, but for *large* values of the time t, $e^{-\lambda_B t}$ can be neglected compared with $e^{-\lambda_A t}$. From Eq. 2-6 we thus have, for large values of t,

$$Q = \frac{\lambda_A}{\lambda_B - \lambda_A} P_0 e^{-\lambda_A t},$$

so that

$$\left(1 - \frac{\lambda_A}{\lambda_B}\right)\lambda_B Q = \lambda_A P,$$

or

$$\frac{dN_B}{dt} = \frac{\lambda_B}{\lambda_B - \lambda_A}\frac{dN_A}{dt}.$$

For large values of t the daughter decays with the half life of the parent, but its activity is *greater* than that of the parent by the factor $\lambda_B/(\lambda_B - \lambda_A)$. This kind of equilibrium is known as *transient* equilibrium. An example of such a case is shown in Fig. 2-7 for barium-140 ($t_{1/2} = 12.74$ days),

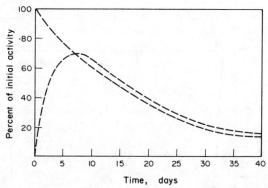

FIG. 2-7   Growth and decay of lanthanum-140 ($t_{1/2} = 40.272$h) from freshly separated barium-140 ($t_{1/2} = 12.74$d) and the decay of the barium-140 as a function of time. The lanthanum-140 activity is a maximum at $t_m = 6.57$d.

which decays to lanthanum-140 ($t_{1/2} = 40.272$ hours). The condition of transient equilibrium gradually changes into that of secular equilibrium as λ_B becomes larger compared with λ_A.

The third case of interest is for a long-lived daughter and a short-lived parent. For sufficiently large values of t, and as $\lambda_A > \lambda_B$, Eq. 2-6 reduces to

$$Q = \frac{\lambda_A}{\lambda_A - \lambda_B}P_0 e^{-\lambda_B t}.$$

In other words the number of atoms Q of element B decay with a half life, or decay constant, characteristic of the daughter nucleus, while the number of atoms P of element A decreases more rapidly with its shorter half life. This situation is illustrated in Fig. 2-8 for the decay of xenon-123 ($t_{1/2}$ = 2.08 hours) to iodine-123 ($t_{1/2}$ = 13.2 hours).

With the artificial production of the transuranic elements, a fourth radioactive family of elements, known as the neptunium series, was discovered. This is shown in Fig. 2-9.

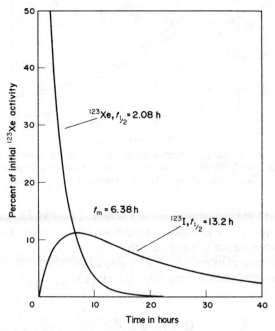

FIG. 2-8   Growth and decay of iodine-123 ($t_{1/2}$ = 13.2h) from xenon-123 ($t_{1/2}$ = 2.08h), and its decay, as functions of time. The iodine-123 activity is a maximum at t_m = 6.6h. Accelerator produced xenon-123 is often used for the production of iodine-123 for medical use.

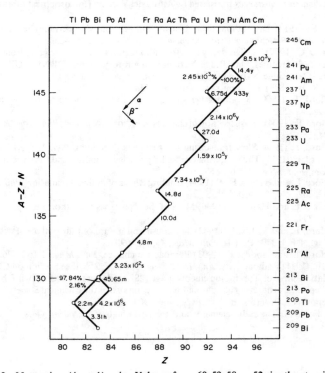

FIG. 2-9 Neptunium (4n + 1) series. Values of n = 60, 59, 58, .. 52 give the atomic-mass numbers of this series. As ^{237}Np is the longest-lived member of this series, its daughters will be in secular equilibrium with it. The neptunium series is now known to be also fed by ^{237}U (β^-, 6.75d) directly to ^{237}Np, and by ^{245}Cm (α, 8500y) through ^{241}Pu (β^-, 14.4y) to ^{241}Am. See D. C. Kocher, 1977.

References

Curie, P. and Curie, M.-P. (1899) Sur la radioactivité provoquée par les rayons de Becquerel, *Comptes Rendus, Académie des Sciences, 129,* 714.

Curie, Sklodowska (1903) Recherches sur les substances radio-actives, *Ann. de Chimie et de Physique 30* [7], 99 and 289. Thesis of Marie Curie that is also available in English as *Radioactive Substances* translated by Alfred del Vecchio (Philosophical Library, New York, 1961).

Fajans, K. (1913) Die radioaktiven Umwandlungen and das periodische System der Elemente, *Berichte der Deutsche chemische Gesellschaft, 46,* 422.

Kocher, D. C. (1977) Nuclear decay data for radionuclides in routine releases from nuclear fuel cycle facilities, Oak Ridge National Laboratory Report ORNL/NUREG/TM-102 (Oak Ridge National Laboratory, Oak Ridge, Tennessee).

Owens, R. B. (1899) Thorium radiation, *Phil. Mag., 48* [5], 360.

Rutherford, E. (1900) A radioactive substance emitted by thorium compounds, *Phil. Mag., 49* [5], 1.

Rutherford, E. (1904) The succession of changes in radioactive bodies, *Phil. Trans. R. Soc.,* A, *204,* 169.

Rutherford, E. (1904) Slow transformation products of radium, *Phil. Mag., 8* [6], 636.

Rutherford, E. (1911) The scattering of α and β particles by matter and the structure of the atom, *Phil. Mag., 21* [6], 669.

Rutherford, E. and Barnes, H. T. (1904) Heating effects of radium emanation, *Phil. Mag., 7* [6], 202.

Rutherford, E. and Hahn, O. (1906) Mass of the α particles from thorium, *Phil. Mag., 12* [6], 371.

Rutherford, E. and Soddy, F. (1902) The cause and nature of radioactivity, Part I, *Phil. Mag., 4* [6], 370; Part II, *Phil. Mag., 4* [6], 569.

Rutherford, E. and Soddy, F. (1903) Radioactive change, *Phil. Mag., 5* [6], 576.

Soddy, F. (1911) Radioactivity, *Ann. Rpts. Prog. Chem.,* 7, 256. (See Section on Chemical relationships of the radioelements page 285.) (Professor Soddy, with but four exceptions, wrote every annual Radioactivity report for the Chemical Society's Annual Reports on Progress in Chemistry from 1904 through 1920.)

Soddy, F. (1913) The radio-elements and the periodic law, *Chemical News, 107,* 97.

3 The Interactions of α, β, and γ Rays with Matter

Up till now it has not been necessary to pay much attention to the energies involved in radioactive transformations. We have briefly referred to the velocities of α and β particles and also to the ranges in air of α particles, an early "rule of thumb" to characterize the penetrating powers of α particles, and hence, implicitly, their energies. Before proceeding further it is important to define the units of energy used in discussing nuclear interactions.

In radioactivity, energy is usually measured in terms of the electron volt (eV), which is the energy acquired when an electron, or any particle having a charge equal to that of the electron, is accelerated through a difference of potential equal to one volt. The relationship between this unit of energy and the SI unit of energy, the joule, is that 1 eV is equal to $1.6021892 \times 10^{-19}$ J. In general, the energies involved in the extra-nuclear-electron transitions giving rise to visible radiation, and in ionizing atoms, are of the order of electron volts or of tens of electron volts. Thus the well-known sodium "D" line corresponds to an approximately 2-electron-volt transition of the electrons involved. Transitions of electrons giving rise to x rays vary from some fifty electron volts to just over one hundred kiloelectron volts ($1 \text{ keV} = 10^3 \text{ eV}$). The energies of α particles and of β and γ rays are normally greater and are measured in terms of kiloelectron volts and megaelectron volts ($1 \text{ MeV} = 10^6 \text{ eV}$).

As has been mentioned before, the radiations from radioactive substances produce (as do the x rays) ionization in air, and it is upon this fact that the electrical method of detection of such radiations is based. In producing such ionization, energy is expended in separating electrons from their parent atoms. For every atom ionized in air, some 35 eV of

37

work must be done. Thus an α particle, in its passage through air, expends some 35 eV of its energy, on an average, for every ion pair that it creates. In creating, say, 10^5 ion pairs it will expend approximately 3.5 MeV of its energy. If this happened to be its initial energy, it would thus be brought essentially to rest, with its power to ionize totally expended.

It is therefore clear, bearing also in mind the homogeneity of the α-particle energies from different α-particle emitters, that for a given density of air (normally at atmospheric pressure and ambient temperature) an α particle of a given energy travels a fairly well-defined distance before expending all but its normal thermal (kinetic) energy and thereby ceasing to be able to ionize the molecules of air. This distance is what was earlier called the *range* of the α particle. It is probably now a somewhat outmoded concept, but it is still often used, just as lead-210 and bismuth-210 continue to be called, respectively, radium D and radium E!

The range of an α particle in air is only "fairly well defined" because the transfer of energy in the creation of ion pairs in air of a given density is a statistical or random process. Thus in the approximately 10^5 collisions of a 3.5-MeV particle before "coming to rest," some may result in the transfer of energy much less than 35 eV and some in more. The collisions with the molecules of air will be of varied kinds, some just removing outer electrons and others giving rise to considerably larger losses of energy. In some cases the electrons ejected from the air molecules may themselves have energies of the order of 1 keV and can themselves cause secondary ionization. In a given distance the numbers of collisions between particles and air molecules will also vary, so that the range, or the distance traveled, will vary statistically around an average. The resulting spread in the range is known as the *straggling* of the α particles.

The earliest measurements of α-particle ranges were made by W. H. Bragg and R. Kleeman, and also by H. Geiger and J. M. Nuttall. Bragg used a shallow ionization chamber with one electrode consisting of a wire grid to measure the specific* ionization (i.e., the ionization per unit length of path) at different distances from the source of a collimated beam of α particles. His apparatus is shown schematically in Fig. 3-1,

---

* "Specific" now generally refers to "per unit mass."

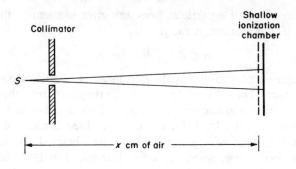

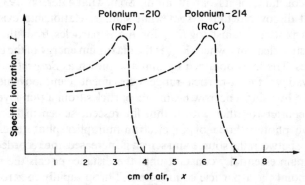

FIG. 3-1   Schematic and results of Bragg's experiment to measure the specific ionization of α particles. (After I. Curie.)

together with the form of the ionization curves that he obtained. The specific ionization increases to a maximum toward the end of the range, and the general form of these curves is usually known as a *Bragg curve*. The experiments of Geiger and Nuttall utilized a spherical ionization chamber with the α-particle source at the center of the sphere. Starting at low pressures the ionization current increased with the pressure of the air, or other gas used, until a pressure was reached at which the range of the α particles was equal to the radius of the sphere. Thereafter the ionization current would remain constant with increasing pressure.

As a result of ionization experiments of this kind carried out in 1910 and 1911, a law known as the Geiger-Nuttall rule was established relating the range of α particles from any given element to the decay constant of that element, namely,

$$\log R = A \log \lambda + B, \tag{3-1}$$

where R and λ are the range and decay constant, respectively, and A and B are constants. The value of A is practically the same for all three of the series of naturally radioactive elements (see Figs. 2-3, 2-4, and 2-5). On plotting $\log \lambda$ against $\log R$ for each of the three families of natural radioelements, three closely parallel lines are obtained. These results thus confirmed a suggestion made by Rutherford in 1907 that there might be a relation between the range of an α particle and the half life, or decay constant, of its parent atom. Somewhat later, in 1933, B. W. Sargent discovered that a somewhat similar relationship exists when $\log \lambda$ is plotted against $\log E_{max}$ for the β particles from the natural radioactive elements, where E_{max} is the maximum energy of the emitted β particles. The lines obtained by him are known as *Sargent curves*.

Nowadays the α-particle-range experiment can conveniently be repeated by using a narrow beam of α particles from a thin layer of any one α emitter together with a thin fluorescent screen in front of the sensitive photocathode of an electron-multiplier phototube. As the distance between the source and screen is increased, the recorded counts will remain essentially constant until the distance equals the range, at which point the α-particle count rate will drop rapidly to zero. Cloud-chamber photographs also spectacularly demonstrate the homogeneity of the range or energy of the α particles from a given radioelement. In such photographs the secondary tracks due to ionization caused by high-energy electrons ejected by an α particle from the gas molecules can also be clearly seen in the form of the so-called δ rays.

Geiger also carried out experiments in Rutherford's laboratory to determine the relationship between the initial velocity V of an α particle and its range. This he did by allowing a collimated beam of α particles from radium C' to be deflected by a magnet and then to be detected by a zinc sulphide screen. The α particles traveled in a vacuum, and their velocities were varied by allowing them to be slowed down by various thicknesses of mica (of known air equivalent range x) before entering the magnetic field. From their deflection in the magnetic field Geiger determined their velocity V, from the relation $V = HEr/M$ (where H is the magnetic field strength, E the charge on the α particle, r its radius of curvature in the magnetic field, and M its mass).

Geiger found that

$$V^3 = a(R - x) \tag{3-2}$$

where a is a constant and $R - x$ is effectively the *residual* range, or the distance the α particle will still travel before it ceases to ionize. For $x = 0$, V is the initial velocity of the α particle of range R cm in air at STP.

Actually the general shape of the Bragg curves was very simply derived from Eq. 3-2 by Geiger by making the plausible assumption that the specific ionization (I) produced by an α particle in its passage through matter is proportional to the energy absorbed, i.e.,

$$I \propto -(d/dx)(\tfrac{1}{2}MV^2),$$

or

$$I \propto -(d/dx)(R - x)^{2/3},$$

whence

$$I \propto (R - x)^{-1/3}.$$

This would give a curve for specific ionization (I) *versus* distance traveled (x) similar to that shown in Fig. 3-2, where as x tends to R, I

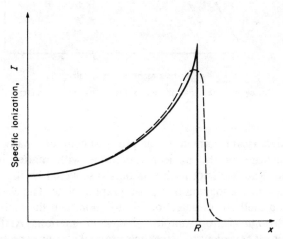

FIG. 3-2   Specific ionization I as a function of distance traveled, x, by the particle. $R - x$ is the *residual* range, the distance left for the particles to travel before coming to rest.

tends to infinity. However, we also know that at $x = R$, V is zero. Thus the particle can no longer ionize so that I must fall rapidly to zero. We therefore obtain a curve which is sharply peaked at $x = R$. However, the straggling of α particles as they pass through matter causes a broadening of the theoretical curve when we consider a large number of α particles instead of just one. Thus we obtain the dotted Bragg curve shown. By such experiments as Geiger's in which the range is determined as a function of the initial energy of the α particles it is possible to plot an α-particle-energy *versus* range curve. Such a curve is shown, quite approximately, in Fig. 3-3.

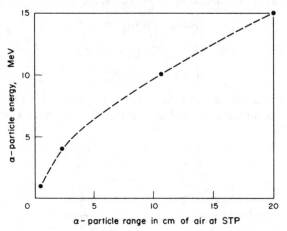

FIG. 3-3 α-particle energies as a function of their range in centimeters of air at STP.

In the historical treatment in the previous chapters, many aspects of the interactions of the nuclear radiations with matter have been considered. Radioactivity itself was only discovered by its interaction with matter, in the fogging of a photographic plate. The electrical and scintillation methods of detection of radiation, and the scattering of α particles by nuclei are examples of such interactions. At this point, however, it is convenient to leave the historical account and to consider the general properties of each kind of radiation.

Essentially the corpuscular or electromagnetic radiation from nuclei, whether it be α, β, or γ, interacts with matter by *imparting energy* to atoms and thereby raising extranuclear electrons to higher energy levels within the atom (excitation) or removing the electron entirely from the atom (ionization). Eventually all of the energy given up in ionizing events is dissipated in atomic or molecular excitation. In addition, γ radiation can lose energy by the creation of electron-positron pairs. The interactions of α and β particles can best be studied in gases, while the methods of study of interactions of electromagnetic radiation depend upon the photon energy. Low-energy photons may conveniently be allowed to interact with the heavier rare gases while high-energy γ rays eject photoelectrons from metal foils and also give readily detectable effects due to photoelectrons, recoil electrons, and electron pairs in, say, a sodium iodide crystal.

When gases are ionized, an electric current can be made to flow through the gas. In turn, this electric current can, under appropriate conditions, be made to serve as a measure of the intensity of the radiation (i.e., the numbers of α, β, or x rays, or the amount of energy that they transfer to the gas by the creation of "ion pairs" of positive ions and electrons).

The α particles are homogeneous in energy, having discrete ranges, and they ionize a gas in a manner that is exemplified by the Bragg curves of Figs. 3-1 and 3-2. The α particle removes electrons from atoms without suffering any appreciable deflection from its rectilinear path provided that it does not approach closely to the nucleus of an atom. Moreover, the α particle may impart enough energy to these electrons that they in turn can ionize the gas, giving rise to the so-called δ rays. Only occasionally, when an α particle penetrates close to the field of a nucleus, is it scattered. The tracks of α particles can be observed in the cloud chamber and also in photographic emulsions that have been exposed to α radiation, developed, and then viewed under moderate magnification. The δ ray appear as small spurs on the α-particle track in a cloud-chamber photograph. The ionization caused by an α particle in, say, air at STP is relatively intense, a 5.5-MeV α particle losing all this energy in about 4 cm of path (see Fig. 3-3).

It has been found experimentally that both α and β particles expend, on an average, between 34 and 35 eV of energy per ion pair created in air.

This quantity is usually designated as W_{air}. As the first ionization potentials of both oxygen and nitrogen are around 14 volts, corresponding to an expenditure of energy of only 14 eV, we see that, *on an average*, another 20 eV are available (per ion pair) to provide residual kinetic energy for the electrons, for excitation of atoms and, lastly, for heating effects.

In this Chapter we have already referred to Geiger's empirical law (Eq. 3-2) relating the velocity of an α particle to its residual range, namely,

$$V^3 = a(R - x),\qquad(3\text{-}2)$$

whence we see that

$$\tfrac{1}{2}MV^2 \propto (R - x)^{2/3}.$$

If we designate the kinetic energy of the α particle by T_α, then the decrease in kinetic energy with distance x is given by

$$dT_\alpha/dx \propto -(R - x)^{-1/3};$$

or, using Eq. 3-2,

$$dT_\alpha/dx \propto 1/V.\qquad(3\text{-}3)$$

Thus as V decreases, the loss of kinetic energy in unit path length due to ionization of air or any other gas (i.e., the specific ionization) increases, in conformity with the shape of the Bragg curve.

Experimentally it is now known that Geiger's formula (Eq. 3-2) holds only approximately between ranges in air at STP of about 3 and 7 cm. Below an energy corresponding to a range of 3 cm (~ 4.5 MeV) the total range (i.e., $x = 0$ in Eq. 3-2) is more nearly proportional to $V^{3/2}$, while above an energy corresponding to a range of 7 cm (~ 7.8 MeV) the range is more nearly proportional to V^4. For these higher energies, therefore, the simple calculation giving relation 3-3, starting with V^4 in Eq. 3-2 instead of V^3, would give

$$\frac{dT_\alpha}{dx} \propto \frac{1}{V^2}.\qquad(3\text{-}4)$$

Another important factor in the interaction of α particles with matter is that, below an energy of a few MeV, an α particle is not always doubly ionized. It can pick up one or two electrons, and it can thus exist part of the time as a singly charged ion and part as a neutral atom. Thus, its ionizing power and its ability to lose energy decrease, and this rather unsteady state of affairs contributes to the straggling, to which reference has already been made.

Relations (3-3) and (3-4) show that the loss of energy per unit path of an α particle (i.e., its specific ionization) is inversely proportional to V or V^2 above energies of a few MeV.

In 1913 Bohr, using classical mechanics and the Rutherford concept of the atom, derived an expression for the energy given up by any charged particle when it passes within a certain distance of an atom. He considered the momentum imparted to an atomic electron, in a direction normal to the path of the ionizing charged particle (see Fig. 3-4). The

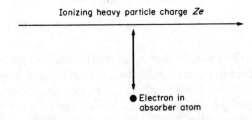

Ionizing heavy particle charge Ze

● Electron in absorber atom

FIG. 3-4   The net momentum of the electron will be perpendicular to the path of the ionizing particle. The component of momentum parallel to the path is zero since the change in momentum as the ionizing particle approaches the atom is equal and opposite to that as it leaves, since it is assumed that the velocity of the ionizing particle is much greater than that of the orbital electron.

expression derived by Bohr showed that dT/dx is proportional to $(1/V^2)\ln kV^3$, where k is a constant involving, among other quantities, the average ionization potential of the atoms in the absorber. In 1915 Bohr modified this expression to include β particles having velocities approaching that of light. In 1930 H. A. Bethe derived, quantum mechanically, a formula for the average rate of loss of energy by inelastic collisions, with distance. This formula, which was valid for high-velocity incident charged particles, involved a change from V^3 to V^2 in the logarithmic term of the classical formula of Bohr.

Relativistic corrections derived in 1932 by both Bethe and C. Møller added the terms $-\ln(1 - V^2/c^2) - V^2/c^2$ to the logarithmic term, c being the velocity of light. In 1933 F. Bloch derived an expression for the average value of dT/dx by inelastic collisions for which the Bethe and classical Bohr formulae were limiting cases for high and low velocities, respectively, of the incident charged particles. In every case, however, the value of dT/dx was proportional to $1/V^2$ multiplied by a factor involving the logarithm of V^2 plus other terms. In general, this logarithmic factor does not have much influence on the $1/V^2$ variation.

We will not consider this matter in great detail; it is merely intended to give an indication that, on theoretical grounds, the proportionalities 3-3 and 3-4 hold equally for α particles and β particles in their passage through matter. The theoretical calculations show that

$$dT_\alpha/dx = -(4\pi e^4 z^2 NZ/mV)\ln k_\alpha \text{ for } \alpha \text{ particles, and}$$

$$dT_\beta/dx = -(4\pi e^4 NZ/mV^2)\ln k_\beta \text{ for } \beta \text{ particles, where}$$

V and z are the velocity and charge number of the incident particle, and N is the number of atoms per cm^3 of the absorber having an atomic number Z. The average values of $\ln k_\alpha$ and $\ln k_\beta$ are not too different, and the expression for dT_α/dx contains an additional factor of 4 for the charge number, z, of the α particle. The former relation for dT/dx holds equally for protons or any light high-velocity charged nucleus of atomic number z and velocity V. For further information and references, the reader should consult the review articles published in 1937 by M. S. Livingston and H. A. Bethe, and in 1953 by Bethe and J. Ashkin.

Thus from these relations for dT_α/dx and dT_β/dx, and because the mass of the α particle is some 7000 times greater than that of the β particle, it is to be expected that the ranges of β particles in matter will be considerably greater than those of α particles of the same energy. In actual fact, dT_β/dx for 4-MeV electrons in air at STP, as calculated by Bethe's relativistic and quantum-mechanical formula, is found to be about 2.2×10^3 eV per cm, while the corresponding value of dT_α/dx, in air at STP, for 4-MeV α particles is about 1.44×10^6 eV per cm.* The calculated value for dT_β/dx for 10-keV electrons is about 2.4×10^4 eV per cm of air at STP.

There are, however, other important differences between the interactions of α and β particles with matter. The former usually has a rectilinear path and a fairly well defined range and suffers only an occasional deflection by too close an approach to an atomic nucleus. It divests itself of its energy chiefly by ionizing the medium through which it passes.

The α and β particles have these points in common: they obey the Rutherford law for nuclear scattering (provided that the β-particle velocity is nonrelativistic, i.e., for relatively low energies), and they both lose energy by creating ion pairs, but in this latter respect the β-particle ranges greatly exceed those of the α particles.

The β particle differs from the α particle in four other important respects:

(1) It suffers many deflections in its encounters with the extra-nuclear electrons, which are "brushed aside" by the heavier particle;

(2) It can create photons, in the form of x rays, when it undergoes drastic acceleration or deceleration in the nuclear field;

(3) It produces different interference phenomena in collisions with identical particles;

(4) It is emitted from the nucleus with energies ranging from 0 to $E_{\beta_{max}}$, whereas different groups of α particles are homogeneous in energy.

The difference in scattering of the two particles (item 1 above) is somewhat analogous to that which we expect between the firing of an artillery shell and a small-caliber bullet into a forest of fairly widely-spaced trees with many branches. The shell would travel in a straight line, tearing off the branches, and quickly be brought to rest; the bullet, on the other hand, would ricochet from branch to branch and frequently change direction. Only when the shell hit the trunk of a tree might it be seriously deflected.

---

* This value of dT_α/dx for a 4-MeV α particle corresponds, assuming that it remains the same at lower energies, to a range of roughly 4/1.44 cm in air at STP. This value is in reasonable agreement with the α-particle-energy *versus* range curve in Fig. 3-3.

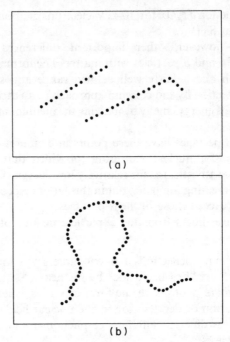

FIG. 3-5   Schematic representation of α-particle tracks (a) and a β-particle track (b) in photographic emulsions.

Thus where the α particles are transmitted through a thin layer of, say, mica largely undeflected and with a definite reduction in energy, the β particles may suffer so many deviations in the material that their path length may be considerably greater than the thickness of the layer, and their original energy spectrum is considerably altered. Thus in a photographic emulsion the paths of α particles might appear as depicted in Fig. 3-5a, whereas that of a β particle would be more like the track illustrated in Fig. 3-5b. A piece of paper of the thickness of this page would stop most of the naturally-occurring α particles, but β particles having energies of 1 MeV can penetrate more than 1 mm of aluminum, and their path lengths in the aluminum can be considerably greater, by virtue of their wanderings, than the thickness of the layer penetrated.

Any charged particle is scattered when it interacts with a Coulomb field either of a nucleus or of an extranuclear electron. The α particles, being a factor of about 7000 greater in mass, do not "notice" the extranuclear electrons and thus suffer scattering only by the nuclei. The electrons, however, being of the same mass, are frequently scattered by the extranuclear electrons in addition to being scattered by nuclei in their passage through matter. When a beam of electrons is incident upon a thick metal sheet, some 10 per cent (for low Z) to 90 per cent (for high Z) are observed to be back-scattered. For the scattering of electrons of low (nonrelativistic) velocity by nuclei the Rutherford scattering law should still hold, with Ze^2 substituted for ZeE. It is practically impossible to verify this experimentally, however, because of the multiple scatterings suffered by electrons and because of the screening effect of the extranuclear electrons around the nucleus. The Rutherford law, in its derivation, took no account, however, of the screening effect of the extranuclear electrons nor of any recoil of the nucleus. For the latter reason the Rutherford law could not possibly be applied to the scattering of incident electrons by the extranuclear electrons of the scattering material. For higher incident-electron energies and for relativistic velocities a scattering formula has been derived by N. F. Mott, but this, for the reasons given above, is difficult to verify by experiment.

On classical theory, when a charged particle is accelerated it radiates electromagnetic energy (item 2 above). Thus when an electron, whether it be a cathode ray or a β particle, is incident upon an absorber and undergoes a sequence of accelerations, it radiates electromagnetic energy in the form of photons. Classically one can see that if a charge at rest, say, is suddenly accelerated, a disturbance must travel outward in space with the velocity of light as the lines of electric force emanating from the charge at rest have to adjust to its new condition of motion. Likewise *on deceleration* a similar electromagnetic disturbance is radiated *outward*. Thus, as was discovered by Röntgen (Chapter 1), cathode rays impinging on a solid give rise to x rays. Such x rays, as distinct from characteristic x rays which were investigated by Barkla and Moseley, form a continuous spectrum *with an upper energy limit equal to the maximum electron energy*. Because these x rays which have a continuous photon energy spectrum arise from the stopping of electrons, they have also been called *bremsstrahlung* (from the German "brake radiation").

Classical theory shows that the amplitude of the radiated electromagnetic energy is proportional to the acceleration of the charge when its velocity is changed or when it is deflected in a collision with another charged particle, as in Rutherford scattering. The intensity of the radiation is also proportional to the square of its amplitude. The electrostatic force at any distance d between the nuclear charge Ze and the α particle, or any other particle, with charge E, is ZeE/d^2. Then, from Newton's law that force is equal to mass times acceleration, the acceleration is ZeE/Md^2 where M is the mass of the particle. The intensity of the electromagnetic radiation arising from the acceleration is thus proportional to $(ZeE/Md^2)^2$ and is therefore proportional to $1/M^2$. Because the mass of the α particle is some 7000 times greater than the mass of the electron, the intensity of electromagnetic radiation produced by an α particle is of the order of a factor of 5×10^7 less (i.e., 7000^2) for the same acceleration and is therefore negligible compared with the bremsstrahlung from electrons having similar energies.

H. A. Kramers in 1923 and G. Wentzel in 1924 investigated this question from the point of view of the quantum theory, as did also H. A. Bethe and W. Heitler in 1934.

It is now necessary to consider briefly the production of the characteristic x rays of the different elements. On Planck's quantum theory, electromagnetic radiation, of wavelength λ and frequency ν, transfers energy in the form of quanta, each quantum (or package of energy) having an energy equal to $h\nu$, where h is Planck's constant of "action" ($h = 6.626 \times 10^{-27}$ erg sec).* Thus in Bohr's theory of atomic spectra, if an atom, consisting of a positive nucleus and extranuclear electrons, changes from an energy level E_1 to an energy level E_2, it will emit a photon of light of frequency ν, such that

$$h\nu = E_1 - E_2. \tag{3-5}$$

In order to account for the discrete lines of optical spectra, Bohr postulated that only certain discrete energy levels of the atom are permitted. The difference between two such energy levels, E_1 and E_2, is

* $h = 6.626176 \times 10^{-34}$ J·s in the SI system.

therefore equal to the decrease in the potential energy of the atom from E_1 to E_2 as an extranuclear electron drops from a higher to a lower potential energy level. Bohr therefore postulated that the extranuclear electrons of the Rutherford atom could only occupy certain orbits, or shells, around the nucleus. These shells of electrons, proceeding out from the nucleus, were called the K, L, M, N, etc., shells. Let us suppose that *an atom*, such as hydrogen, with an electron in the K shell *has a potential energy* W_K. If the atom receives sufficient external energy to send the electron out to the L shell (with a higher potential energy W_L), then, if the electron drops back to the K shell, a photon of energy, $hv = W_L - W_K$, is radiated as the potential energy of the atom decreases from W_L to W_K. In the hydrogen atom the energy transition $W_I - W_K$ is small enough to correspond to the frequencies of the visible spectrum. For heavier atoms, as Z increases, the binding force between the nucleus and the K-shell electrons increases (and, of course, other electrons are added to the L, M, and outer shells). Thus for heavy atoms the transition energy $W_L - W_K$ increases, and consequently the frequency v is much greater and the corresponding wavelength ($\lambda = c/v$) is much shorter. As Z increases, λ for the transition W_L to W_K decreases toward the ultraviolet and then to the x-ray regions of the electromagnetic spectrum. Thus the *characteristic* x rays arise from such transitions to the K shell (or the L shell, etc.) in the heavier elements.* Moreover (as we had noted earlier in the work of Moseley), the frequencies of these characteristic x rays ($hv_{KL} = W_L - W_K$) must also increase consecutively from element to element as the atomic number Z increases.

   Thus, in brief, the Bohr theory postulated a number of stable *stationary states* of the atom, and it also tried to depict these as consisting of electrons in different elliptical orbits around the nucleus. When an electron drops from an outer to an inner orbit, the system of the nucleus plus electron (i.e., the atom) decreases in potential energy, and energy is radiated as a photon.

   On the quantum theory the frequencies of the x rays of the bremsstrahlung spectrum are simply derived from the equation

$$hv = W_i - W_f, \tag{3-6}$$

---

* Characteristic K x rays are of the order of 1 keV in energy at $Z = 11$ (i.e., sodium).

where W_i and W_f are the initial and final energies of the decelerated electron (W_i is equal to eV where V is the accelerating voltage, or eV is the energy of the β particle). If the electron loses all its energy in the collision, W_f is zero and the maximum frequency (or energy) of the bremsstrahlung spectrum can be deduced from the equation

$$hv_{max} = W_i = eV. \tag{3-7}$$

The average rate of radiation energy loss for electrons having very large values of the kinetic energy $T(T \gg 137mc^2/Z^{1/3})$ was calculated by Bethe and Heitler in 1934 to be given by

$$\left(\frac{dT}{dx}\right)_{rad} = -\frac{Z^2 N r_0^2 T}{137}\left(4\ln\frac{183}{Z} + \frac{2}{9}\right), \tag{3-8}$$

where r_0 is the "classical" radius of the electron ($r_0 = e^2/mc^2$) and the constant 137 is the calculated value of $hc/2\pi e^2$. For lower values of $T(mc^2 \ll T \ll 137mc^2/Z^{1/3})$, Heitler and F. Sauter found that

$$\left(\frac{dT}{dx}\right)_{rad} = -\frac{Z^2 N r_0^2 T}{137}\left(4\ln\frac{2T}{mc^2} - \frac{4}{3}\right). \tag{3-9}$$

Bethe and Heitler also found a very simple and approximate relation for the relative rates of loss of energy by electrons by the processes of collision (i.e., ionization and atomic excitation) and radiation in material of atomic number Z, namely, that

$$\frac{\left(\frac{dT}{dx}\right)_{coll}}{\left(\frac{dT}{dx}\right)_{rad}} \sim \frac{1600mc^2}{TZ} \tag{3-10}$$

The energy, T_c, at which the rate of loss of energy by radiation equals that by the process of collision (i.e., when the right-hand side of Eq. 3-10 is unity) is given by

$$T_c = 1600mc^2/Z \tag{3-11}$$

For electrons in lead the value of T_c at which the average rates of radiative and collision energy losses become equal is approximately

T. For cathode rays having energies less than a few hundred kiloelectron volts, so small a proportion of the energy produces x rays and so large a proportion produces heat, by ionization, that it is necessary to cool the targets of low-voltage x-ray tubes. At higher energies such cooling becomes unnecessary.

Having considered items 1 and 2 of our list of the properties of β particles that differ from those of α particles, let us now pass to items 3 and 4.

Item 3 is concerned with the interference effects in collisions between identical particles. This problem was considered in detail by N. F. Mott in 1929 and 1930 in his treatment of the problem of the scattering of electrons by atoms. The Schrödinger wave representing the incident β particle and the Schrödinger wave representing the recoiling electron 10 MeV, while for electrons in air T_c is about 100 MeV. In Fig. 3-6 are indicated graphically the variation of the average rates of energy loss by radiation and by collision as a function of energy for electrons in both lead and air. The rate of energy loss is not too greatly different at lower energies for materials as different as air and lead. These curves show that the electron energy loss as a result of x-ray production is very inefficient below 1 MeV compared with electron energy losses by ionization and excitation. This is also apparent from the approximate relation 3-10 where it is seen that the ratio of energy loss producing x rays to that causing ionization and excitation is proportional to the electron energy can produce interference effects at certain angles, in just the same way that two beams of light can produce interference effects. (Similar interference effects arise in the scattering of α particles by helium or protons by hydrogen. The diffraction effects observed by Davisson and Germer and by G. P. Thomson with beams of electrons, while demonstrating the wave nature of the electron, are different in that they arise from the scattering of the electrons, by the atoms of the crystal lattice, without loss of energy *hv* or change in wavelength *λ* of the scattered electrons. This is called *coherent* scattering.) The relations for *dT/dx* derived by Bethe and his associates have taken account of this interference effect in the collisions between electrons.

Item 4 concerns the shape of the β-ray spectrum. As early as the turn of the nineteenth century Becquerel had shown that the energies of the β particles emitted by any given radioactive element varied continuously

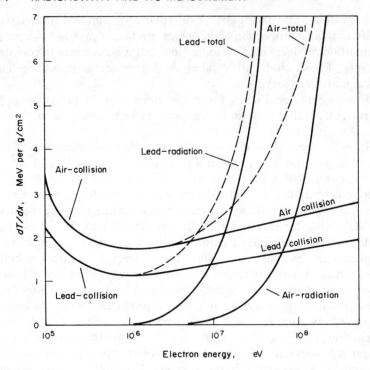

FIG. 3-6   Average rates of energy loss for electrons, in air at STP and in lead, by collision and radiation as a function of electron energy. Collision includes ionization and atomic excitation. There are small differences of dT/dx for both collision and radiation between positive and negative electrons.

in energy. The method he used, which is illustrated in Fig. 1-2, contained the principle of a present-day 180° β-ray spectrometer. A chief difference is, however, that nowadays, instead of a photographic plate, one normally uses a β-particle detector behind a narrow slit and "sweeps through" the β-ray spectrum by increasing or decreasing the magnetic field. Various forms of β-ray spectrometer (including the use also of electrostatic deflection), with total deflections through different angles, have also been used.

An example of a β-ray spectrum is shown in Fig. 3-7, for carbon-14, which emits β particles without any associated γ radiation.

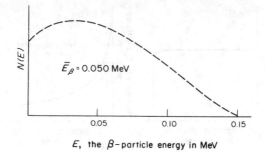

$$\bar{E}_\beta = 0.050 \text{ MeV}$$

E, the β-particle energy in MeV

FIG. 3-7   The β-ray spectrum of carbon-14. $N(E)$ is the number of β particles emitted in a small fixed energy range ΔE.

We shall return in a later chapter to consider the shape of the β-ray spectra. It is, however, now apparent that the absorption of β particles, because of their different processes of giving up energy, their more frequent scattering, and their continuous energy spectra of emission from nuclei, are far more complicated than the absorption of α particles. Fortunately, however, all these different factors combine to give, in practice, an approximately exponential decrease in the number of β particles transmitted by an absorber. Figures 3-8a and 3-8b compare the transmission curves for α and β particles, the latter decreasing roughly exponentially until it merges into the background of secondary electrons due to bremsstrahlung (for a pure β emitter) or bremsstrahlung and γ rays. Figure 3-8c shows a similar β-particle absorption curve plotted semilogarithmically. The range of the β particles in the absorber is often taken as the point of intersection of the β-particle component of the curve with the secondary electron background. Where, however, the range of the α particle is equal to its total path length in the absorber, the path length of the β particle will, as may be seen in Fig. 3-5b, be much greater than its range.

A second fortunate and useful circumstance is that the absorption of β particles is approximately the same for any of the light elements, *provided* that the amount of absorbing material is expressed in terms of *mass per unit area*, e.g., milligrams per square centimeter. (The product of absorber thickness and its density gives the mass per unit area interposed

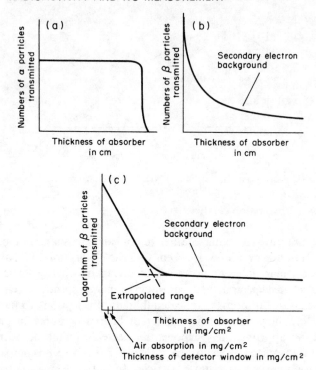

FIG. 3-8   Transmission curves for α and β particles through matter. Such experimentally determined transmission, or absorption, curves used in conjunction with the data shown in Figs. 3-3 and 3-9 permit the energy measurement of α and β particles from simple α and β emitters.

between a radioactive source and a radiation detector.) Thus a Geiger-Müller-counter window of $0.1\,\text{mg/cm}^2$ has very closely the same absorption for β particles of a given energy as an aluminum foil of $0.1\,\text{mg/cm}^2$ or of a "thickness" of air of $0.1\,\text{mg/cm}^2$. Thus, provided that all absorbers are expressed in milligrams per square centimeter, the air or the mica window of a radiation detector or an aluminum absorber can all be plotted additively as abscissae in a semilogarithmic plot of the type illustrated in Fig. 3-8c. For a curve of this type the numbers of β particles transmitted by an absorber is given closely by the relation $I = I_0 e^{-\mu x}$, where I_0 is the initial number for no absorber, x is the thickness of the absorber in, say, centimeters, and μ is then the *linear* absorption coefficient in cm^{-1}. The actual shape of the absorption curve is quite dependent on the source-absorber-detector geometry and the relative efficiencies of the particular detector used, for β particles and for bremsstrahlung. The range, however, as measured in Fig. 3-8c, should be less dependent on the various experimental parameters. To express this approximate exponential attenuation in terms of milligrams, or grams, per square centimeter m_x instead of in distance x in centimeters, we use the relation $x\rho = m_x$, where ρ is the density of the absorber in grams per cubic centimeter. Thus we have

$$I \sim I_0 e^{-(\mu/\rho)m_z}. \tag{3-12}$$

The *mass* absorption coefficient is not only very nearly the same for all low-Z absorbers for the β particles from a given radioactive element, but it increases only slowly with increase in atomic number when the heavier elements are used as absorbers. The value of μ/ρ is also independent of the physical state of the absorber, whether it be gaseous, liquid, or solid, because the absorption of the β particles, resulting from their interactions with the nuclei and extranuclear electrons of *atoms*, depends merely on the number of atoms encountered; thus the same numbers of atoms give the same absorption independent of their aggregation. The foregoing considerations apply to the variation of absorber for a given β-ray spectrum. With increasing β-particle energies, however, μ/ρ decreases. The range of the phosphorus-32 β particles ($E_{\beta_{\max}} = 1.71$ MeV) as determined from an absorption curve (such as that of Fig. 3-8c) is around $750\,\mathrm{mg/cm}^2$ in aluminum and $860\,\mathrm{mg/cm}^2$ in lead.

In 1931 N. Feather proposed an empirical rule to relate the maximum β-ray energy with the range R in an absorber. This rule, which holds approximately for β particles of maximum energy above about 0.8 MeV, is that

$$R = 543 E_{\beta_{\max}} - 160, \tag{3-13}$$

where R is expressed in milligrams per square centimeter and $E_{\beta_{\max}}$ is in million electron volts of energy. E. C. Widdowson and F. C. Champion, in 1938, proposed slightly different constants for the Feather rule, namely, 536 and 165, which seem to fit well with the experimental data above 0.8 MeV. In 1953, W. J. Whitehouse and J. L. Putman proposed the relationship

$$R = 370E_{\beta_{\max}}^{3/2}, \tag{3-14}$$

which they found to agree to within 20 per cent with the experimental data in the range 0.025 to 2 MeV.

From the empirical relations 3-13 and 3-14 the range is seen to vary linearly with the maximum energy for high-energy β-particle spectra,

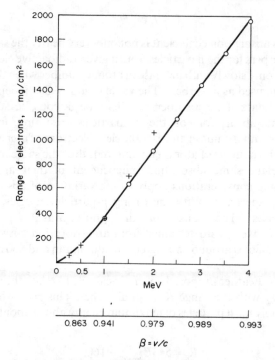

FIG. 3-9   Range of β particles or electrons, in milligrams per square centimeter as a function of their maximum energy, in a low-Z material such as aluminum. The circles are obtained from Eq. 3-13 (Feather rule as modified by Widdowson and Champion) and the crosses (+) from Eq. 3-14 (Whitehouse and Putman). These ranges are extrapolated ranges (Fig. 3-8); actual ranges in the absorber can be much longer (Fig. 3-5). Values of $\beta = v/c$ are also shown as a function of electron energy.

and to the three-halves power at lower energies. The relationship between range and energy is illustrated in Fig. 3-9.

Gamma rays are emitted from the nucleus of an atom when the nucleus reorganizes and makes a transition from a high-energy state to a lower energy state or to the ground (or normal) state of the nucleus. After the emission of an α or β particle, the daughter nucleus may be left in an excited state, i.e., with an excess of energy over the ground state. This surplus energy can be emitted in the form of a γ ray. When carbon-14 and phosphorus-32 decay to nitrogen-14 and sulphur-32, respectively, the whole of the available energy is used in the β transition and the decay is to the ground states of the nitrogen-14 and sulphur-32 nuclei. Thus *no* γ radiation is observed as a result of these transitions.

A γ ray is usually characterized by its energy hv, and so we speak of the 1.173-MeV γ ray in the decay of cobalt-60 which is followed by the 1.332-MeV γ ray "in cascade." These transitions are normally depicted by a decay scheme such as is shown in Fig. 3-10. The β-particle transition is shown as going from left to right in order to conform with the *increase* of one in the atomic number, with the *loss* of one unit of negative charge from the cobalt-60 nucleus. In fact the 1.173-MeV and 1.332-MeV γ rays are emitted by nickel-60 and *not* by cobalt-60; they are, however, often loosely designated as the cobalt-60 γ rays.

The interactions of γ rays with matter are chiefly interactions with electrons which occur in three principal ways:

(1) The photoelectric effect, which is important at relatively low energies;

(2) The Compton effect, which is the scattering of γ rays by electrons; and

(3) The creation of positron-electron pairs (pair production), which is important at γ-ray energies above 1.022 MeV.

In the photoelectric effect (which is exactly parallel to the photoelectric effect with visible or ultraviolet light, but with different hv) the whole of the energy of the γ ray is absorbed in an environmental atom and the energy is then carried away by one of the extranuclear electrons. The energy of the resulting photoelectron is equal to the energy of the γ ray *less* the energy required to remove the photoelectron from its parent atom. Thus, if T_K, T_L, T_M, etc. are the energies of the photoelectrons that originate, respectively, in the K, L, M, etc., shells of the atom, and

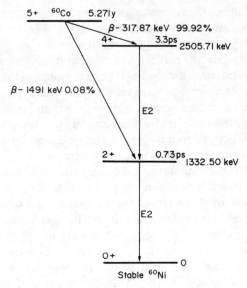

FIG. 3-10 Schematic representation of the decay of ^{60}Co. Nuclear data are being constantly revised and recent nuclear-decay-data compilations should be consulted for the latest information. Nuclear transitions are subject to certain selection rules based on quantum numbers such as nuclear angular momentum (or "spin"), and parity. The numbers 5, 4, and 2 refer to $5h/2\pi$, $4h/2\pi$, and $2h/2\pi$ units of angular momentum. Even parity is designated by "+" and odd by "−". E2 signifies that the electromagnetic radiation emitted in the nuclear transition is electric quadrupole. For a discussion of spin, parity, and multipolarities, which is beyond the scope of this book, the reader is referred to R. D. Evans (1955) or S. A. Moszkowski (1965).

W_K, W_L, W_M, etc. are the respective electron-binding energies, then

$$T_K = h\nu - W_K,$$
$$T_{L_I} = h\nu - W_{L_I}, \qquad (3\text{-}15)$$
$$T_{L_{II}} = h\nu - W_{L_{II}}, \quad \text{etc.}$$

Another mode of radioactive decay by which the daughter nucleus can discard its residual energy is that of *internal conversion*, in which the excitation energy of the nucleus is transferred to one of its extranuclear electrons. In this transition, no γ ray is emitted, although the term "internal conversion" must originally have had the connotation of a γ ray interacting with an electron of the *same* atom. The excess energy of the daughter nucleus corresponding to its state of excitation is simply transferred to one of the extranuclear electrons, which is expelled with kinetic energy equal to the de-excitation energy in the transition from the excited state of the daughter nucleus *less* the electron-binding energy appropriate to the atomic shell from which the electron originates. In the decay of mercury-203, which is illustrated in Fig. 3-11, the 0.28-MeV de-excitation energy of the daughter thallium-203 nucleus is emitted in the form of γ radiation in only 81.5 per cent of the transitions, the remaining 18.5 per cent occurring as internal conversion electrons, or, as they are now usually called, "conversion electrons". The *branching ratio* between these two modes of decay is expressed in terms of the *internal conversion coefficient*, which is the ratio of conversion-electron to gamma-ray transitions. The energies of these electrons are given by Eqs. 3-15, except

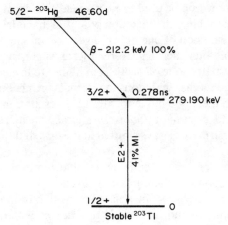

FIG. 3-11   Schematic representation of the decay of ^{203}Hg. Here the nuclear angular momenta of the three nuclear states are $5h/2\pi$, $3h/2\pi$, and $h/2\pi$, and one is of odd and two of even parity. The γ-ray transition is mixed electric quadrupole and magnetic dipole. Internal-conversion coefficients (N_e/N_γ) have values that are dependent upon the difference between the energies of the initial and final states of the transition, the atomic number of the nucleus, the multipolarity of the transition, and the atomic shell from which the conversion electron originates (see R. D. Evans, 1955 and S. A. Moszkowski, 1965). K-shell conversion electrons will leave the atoms with kinetic energies equal to 279.2 keV *less* the binding energy, W_K, of the *thallium* atom, which is about 85.5 keV.

that $h\nu$ now represents the de-excitation energy that *would* have been carried away by a γ ray of energy $h\nu$, if that had been the transition mode. In the deday of cobalt-60, essentially all of the de-excitation energy of the daughter nickel-60 nucleus is carried away in the form of γ radiation.

That the γ-ray and conversion-electron emission actually *follow* that of the α or β particle is confirmed by the fact that values of $W_K - W_L$, etc., are found experimentally to be the differences of energies between consecutive electron shells of the *daughter* atom. Thus the values of $W_K - W_L$, etc., for the experimentally measured energies of the conversion electrons resulting from the decay of iodine-131 (Fig. 3-12) are found to correspond to the differences in the binding energies of the extranuclear electrons of xenon and *not* of iodine. When an electron is ejected from the K, or other, shell of an atom as a conversion electron, x rays can be emitted by the filling of the shell vacancy by another electron. The resulting x-ray energies are also found to be characterstic of the daughter atom and not of the parent. Some of this available "x-ray energy" can, however, be used to eject an electron from an outer shell of the *same* atom that has an ionization potential that is less than the "x-ray energy." The emission of one or more such outer-shell electrons, called *Auger electrons* after their discoverer Pierre Auger, in place of x rays is merely an alternate mode of decay from higher to lower electron-energy levels of excited atoms, analogous to internal conversion. The branching ratios between these two modes of atomic de-excitation are expressed in terms of the *Auger yield* or *fluorescence yield*, which are the fractions of outer-shell electrons or x rays emitted in the de-excitation of the atomic-energy levels.

It cannot be too greatly emphasized that the process of radioactive decay is the response of the *whole atom* to energy that becomes available by virtue of the decrease in mass in the transition from the parent to the daughter atom. This subject, namely the *energetics of nuclear decay*, will be dealt with more fully in Chapter 5.

Figure 3-12 shows the experimentally determined β-ray spectrum of iodine-131. The resulting excited nuclei of xenon-131 emit many γ rays corresponding to different energy levels, which can also decay by emitting conversion electrons. These appear as "spikes" on the continuous β-ray spectrum shown in Fig. 3-12.

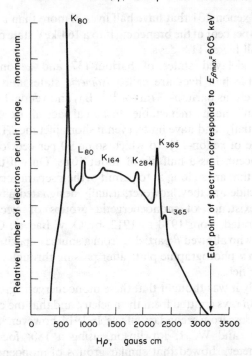

FIG. 3-12   Momentum spectrum of β particles and conversion electrons arising from the decay of iodine-131. The product $H\rho$ of the magnetic field H and the radius of curvature ρ of the path of an electron in that field is proportional to the momentum of the electron. The line spectra correspond to K- or L-shell conversion electrons from the 80-, 164-, 284-, and 365-keV excited states of xenon-131. [After B. D. Kern, A. C. G. Mitchell, and D. J. Zaffarano, *Physical Review*, 76, 94 (1949).]

Normally the γ ray or conversion electron follows the α or β particle in a time that is less than 10^{-12} second (1 ps). Sometimes, however, there is a delay before the γ ray or conversion electron is emitted. Thus cesium-137 decays by the emission of $\alpha\beta^-$ particles to a 662-keV excited state of barium-137, which then decays with a half life of 2.552 minutes to the ground state of barium emitting a 662-keV γ ray, or a conversion electron. More than 90 per cent of the β^- decays of iodine-131 are to two states of xenon-131 that have half lives of more than a picosecond, and some 0.4 per cent of the branching is to a 164-keV state of xenon-131 that has a half life of 11.8 days.

These delayed states of barium-137 and xenon-131 that have measurable half lives are called *isomeric* states and are designated, respectively, as barium-137m (or ^{137m}Ba) and xenon-131m (or ^{131m}Xe), where "m" means metastable. In actual fact all excited states decay exponentially, and have finite, even if short, lifetimes. Thus the 364.480-keV state of xenon-131, to which some 90 per cent of the iodine-131 decays occur, has a half life of about 10 ps. Only if these states have lifetimes that are so long as to permit of their separation from the parent radionuclide are they, however, usually designated as isomeric.

The existence of monoenergetic groups of electrons was first demonstrated, from 1910 to 1912, by O. v. Baeyer, O. Hahn and L. Meitner who allowed β particles, from a source deposited on a fine wire, to fall on a photographic plate after passing through a relatively weak magnetic field.

Initially it was thought that these monoenergetic groups of electrons were the β rays emitted from the nucleus, and that the continuous β-ray spectra arose from secondary effects. In 1914, however, Rutherford, H. R. Robinson and W. R. Rawlinson, using a 180° focussing magnetic spectrometer, showed that similar groups of monoenergetic electrons could be generated by irradiating *external* lead and gold targets with the γ rays emitted in the decay of radium B (lead-214) and radium C (bismuth-214). They therefore concluded that these monoenergetic groups were extranuclear in origin. In 1914, J. Chadwick, again using a 180° focussing magnetic spectrometer, showed that the true β-ray spectrum was continuous in energy, a fact that gave rise to other difficulties that are discussed in Chapter 4. Rutherford and A. N. de C. Andrade, using diffraction by a crystal were also able to show, in 1914, that x rays approximately equal in energy to the K and L x rays of lead were emitted in the decay of radium B. These x rays arising from the filling of atomic-electron-shell vacancies, actually in the atom of bismuth, confirmed the extranuclear origin of the monoenergetic electron groups, which were subsequently called conversion electrons. A general review of these experiments and conclusions was given by Rutherford in 1914 in the *Philosophical Magazine*. Such conversion electrons are observed widely in radioactive decays that leave the daughter nucleus in an excited energy state.

A γ ray cannot give up *all* its energy to a *free* electron because the principles of the conservation of energy and of momentum cannot both be observed. In such a process the W terms in Eq. 3-15 would be zero because the electron is free. Hence $T = h\nu$, i.e., $\frac{1}{2}mv^2 = h\nu$, and the energy of the electron would be uniquely determined, without any reference to the changes in momenta that would be involved. In the case, however, of an electron that is bound in an atom, the atom itself can recoil and the laws of the conservation of energy and momentum can both be preserved.

A γ ray can, however, interact with a free electron and be *scattered* by it in the process known as the Compton effect. It had been known that, in the scattering of x rays, the scattered, or secondary, x rays were less penetrating than the primary beam. In a series of beautiful experiments published in 1923, A. H. Compton showed that the scattered x rays consisted of an *unmodified component* with the same wavelength as the incident x rays and a *modified component* whose wavelength was greater than that of the incident x rays by an amount that depended on the angle of scattering. Compton's experiments were carried out with the K_α x rays from molybdenum, using carbon as the scattering material.

In order to understand the Compton effect it is necessary to use two important results from electromagnetic theory and relativity theory. Electromagnetic radiation transfers energy through space, and when it falls upon a surface it exerts a pressure. This pressure was observed experimentally in 1901 by E. F. Nichols and G. F. Hull at Cornell and by P. Lebedev in Russia. With the energy density of such radiation there must also be associated a momentum density. It can be shown that the momentum density for radiation in free space is equal to the energy density divided by the velocity of light. Thus the momentum of any individual photon is also equal to its energy divided by the velocity of light. Therefore the momentum p of a photon of frequency ν, and hence of energy $h\nu$, is

$$p = h\nu/c. \qquad (3\text{-}16)$$

The second principle that we must use from relativity theory is that the momentum p_e and the kinetic energy T_e of an electron of rest mass m_0

moving with a velocity v, which is comparable with the velocity of light, are given by

$$p = m_0 v / \sqrt{1 - (v/c)^2} \qquad (3\text{-}17)$$

and

$$T = m_0 c^2 \left(\frac{1}{\sqrt{1 - (v/c)^2}} - 1 \right). \qquad (3\text{-}18)$$

Using the binomial theorem to expand the term $(1 - (v/c)^2)^{-1/2}$, we have that

$$(1 - (v/c)^2)^{-1/2} = 1 + \tfrac{1}{2}(v/c)^2 + 3/8(v/c)^4 + \ldots.$$

Substituting this into Eqs. 3-17 and 3-18, we see that, for small values of v where $(v/c)^2$ is so small that it can be neglected compared with unity and $(v/c)^4$ can be neglected compared with $(v/c)^2$, $p = m_0 v$ and $T = \tfrac{1}{2} m_0 v^2$. These are the usual nonrelativistic electron momentum and kinetic energy.

Let us now consider an x-ray photon of energy $h\nu$ which is scattered by an electron, at 0 in Fig. 3-13, through an angle θ with the electron recoiling at an angle ϕ to the initial direction of the incident photon. Let us suppose that the energy of the scattered photon is $h\nu'$ and that the velocity of the recoiling electron is v.

Then from the principle of the conservation of energy and using Eq. 3-18, we see that

$$h\nu = h\nu' + m_0 c^2 \left\{ \frac{1}{\sqrt{1 - (v/c)^2}} - 1 \right\}. \qquad (3\text{-}19)$$

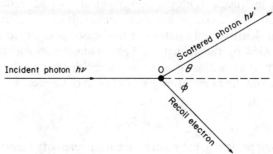

FIG. 3-13 Compton scattering of a photon by a free electron.

From the principle of the conservation of momentum and using Eqs. 3-16 and 3-17, we find that

$$\frac{hv}{c} = \frac{hv'}{c}\cos\theta + \frac{m_0 v}{\sqrt{1 - (v/c)^2}}\cos\phi, \qquad (3\text{-}20)$$

$$0 = \frac{hv'}{c}\sin\theta + \frac{m_0 v}{\sqrt{1 - (v/c)^2}}\sin\phi. \qquad (3\text{-}21)$$

If we take to the left-hand side of Eqs. 3-20 and 3-21 only the terms with ϕ, and if we then square and add, we have

$$\left(\frac{m_0 v}{\sqrt{1 - (v/c)^2}}\right)^2 = \left(\frac{hv}{c}\right)^2 + \left(\frac{hv'}{c}\right)^2 - 2\left(\frac{hv}{c}\right)\left(\frac{hv'}{c}\right)\cos\theta. \quad (3\text{-}22)$$

If we substitute $\eta = 1/\sqrt{1 - (v/c)^2}$ in Eqs. 3-19 and 3-22 and also divide Eq. 3-22 by $m_0^2 c^2$, then from Eq. 3-19

$$hv + m_0 c^2 = hv' + m_0 c^2 \eta \qquad (3\text{-}23)$$

and from Eq. 3-22

$$\left(\frac{hv}{m_0 c^2}\right)^2 + \left(\frac{hv'}{m_0 c^2}\right)^2 - 2\left(\frac{h}{m_0 c^2}\right)^2 vv'\cos\theta = \frac{v^2 \eta^2}{c^2}. \qquad (3\text{-}24)$$

But since $\eta^2 = 1/(1 - (v/c)^2)$, $v^2 \eta^2/c^2 = \eta^2 - 1$. We can therefore rewrite Eq. 3-24 as

$$\left(\frac{hv}{m_0 c^2}\right)^2 + \left(\frac{hv'}{m_0 c^2}\right)^2 - 2\left(\frac{h}{m_0 c^2}\right)^2 vv'\cos\theta = \eta^2 - 1, \qquad (3\text{-}25)$$

whence

$$(hv)^2 + (hv')^2 - 2h^2 vv'\cos\theta + (m_0 c^2)^2 = (m_0 c^2 \eta)^2. \qquad (3\text{-}26)$$

But from Eq. 3-23, $m_0 c^2 \eta = hv - hv' + m_0 c^2$. Squaring this to give $(m_0 c \eta)^2$, substituting in Eq. 3-26 to eliminate η, and cancelling the terms $(hv)^2$, $(hv')^2$, and $(m_0 c)^2$ from each side, we are left with

$$v' = v \left/ \left(1 + \frac{hv}{m_0 c^2}(1 - \cos\theta)\right)\right. . \qquad (3\text{-}27)$$

When θ is zero, $v' = v$ in Eq. 3-27, and the frequency is unchanged. Since, however, the corresponding wavelengths are given by $\lambda = c/v$ and $\lambda' = c/v'$, we have, on substituting for v and v' in Eq. 3-27,

$$\lambda' - \lambda = (h/m_0c)(1 - \cos\theta). \tag{3-28}$$

This formula, derived by A. H. Compton and also by P. Debye, was verified by the experiments carried out by Compton. It shows that the increase of wavelength, $\lambda' - \lambda$, at any given angle of scattering, θ, is completely *independent* of the initial wavelength of the incident photon. The constant h/m_0c, which is equal to 2.426×10^{-12} m, is called the Compton wavelength. Eq. 3-27 shows that the fractional change in v (and hence of the energy hv) is dependent upon the initial quantum energy hv for finite angles of scattering. The larger the energy hv of the incident photon, the larger is the fractional energy drop.

A further point of great interest emerges from Eq. 3-28. If instead of having a free electron the x-ray photon is scattered by an electron that is tightly bound to an atom, then instead of m_0 we must substitute a greater mass, involving the mass of the atom also, in the denominator. The change in wavelength, $\lambda' - \lambda$, is then negligible for all angles, including θ equal to zero, and we have the *coherent* scattering of x rays from atoms which permits x-ray diffraction from crystal lattices. This coherent, or unmodified component of the scattered radiation becomes decreasingly small with increasing frequency, i.e., as the x-ray energy increases.

For the same reason, if we consider high-energy photons, such as γ rays having energies of the order of a million electron volts, then the outer extra-nuclear-electron binding energies are negligible by comparison and the γ rays eject them as the so-called Compton electrons or Compton recoils.

Before considering the phenomenon of the production of electron-positron pairs by γ rays, we must again dwell for a moment upon the simpler relations given by the theory of relativity. It has already been briefly mentioned in Chapter 2 that in some circumstances a radioactive substance emits a positively charged electron, or *positron*. From the historical point of view it is now incumbent upon us to consider briefly the discovery of the positive electron.

The fundamentals of relativity theory must be the subject of a separate study, but we can, as we have earlier in Eqs. 3-17 and 3-18, merely borrow from its simpler relationships. The relativity expressions for the momentum p, the kinetic energy T, and the total energy E of an electron of mass m, moving with a velocity v, but whose mass when stationary (the *rest mass*) is m_0, are

$$p = m_0 v / \sqrt{1 - (v/c)^2}, \tag{3-17}$$

$$T = m_0 c^2 \left(\frac{1}{\sqrt{1 - (v/c)^2}} - 1 \right), \tag{3-18}$$

and
$$E = m_0 c^2 \left(\frac{1}{\sqrt{1 - (v/c)^2}} \right). \tag{3-29}$$

The difference between the total energy E and the kinetic energy T is merely the self-energy $m_0 c^2$ of an electron at rest. The mass at any other velocity v is given simply by the relation $m = m_0 / \sqrt{1 - (v/c)^2}$. From Eqs. 3-17 and 3-29 we can readily derive the relation that

$$E^2 = m_0^2 c^4 + p^2 c^2. \tag{3-30}$$

For an electron at rest for which the momentum p is zero, Eq. 3-30 reduces to

$$E = \pm m_0 c^2. \tag{3-31}$$

In normal practice we only consider the positive solution of Eq. 3-31, to give the well-known relation of Albert Einstein between the self-energy (or inherent energy) of a given mass m_0 at rest.

Both W. Heisenberg and E. Schrödinger had developed a non-relativistic theory to account for the movements of electrons in an electromagnetic field. To account, however, for certain anomalies in the atomic spectra of elements G. E. Uhlenbeck and S. Goudsmit introduced, in 1925, the concept of an electron having spin. Subsequently, W. Pauli and C. G. Darwin incorporated the spin of the electron into the nonrelativistic wave equation and were able to obtain results that agreed with experiment, at least approximately, for spectra such as those of hydrogen.

In 1928 P. A. M. Dirac developed a relativistic wave equation to describe the motions of electrons in an electromagnetic field. He pointed out subsequently, in 1930, that there were two sets of solutions to the relativistic wave equation, the one corresponding to the positive root and the other to the negative root, in deriving the total energy E in Eq. 3-30. There were thus two whole sets of solutions possible, the one corresponding to positive E and the other to *negative E*, i.e., from Eq. 3-30

$$E \geqslant m_0 c^2, \tag{3-32}$$

or $$E \leqslant -m_0 c^2. \tag{3-33}$$

Figure 3-14 illustrates the energy-level diagram for a free electron that would correspond to Eqs. 3-32 and 3-33. Normally on classical theory an electron could not cross the forbidden region between $+m_0 c^2$ and $-m_0 c^2$, but on quantum theory such a jump is possible, just as an electron can drop from an L shell in an atom to the K shell and emit electromagnetic radiation in the process.

The question then arises as to why all electrons do not occupy the states of negative energy instead of existing as free observable electrons with definite amounts of *positive* energy. To meet this difficulty Dirac made the two very fundamental assumptions that:

(1) Nearly all of the negative-energy states in the world are filled, in the Pauli sense; and

(2) The normal electrification of the world (the ground zero potential) corresponds to the zero level in Fig. 3-14 when *all* negative-energy states only are filled.

By "in the Pauli sense" we are referring to the exclusion principle which Pauli put forward in 1925, in order to explain certain features of atomic spectra. This principle says that no two electrons (and, later, no two nucleons) can occupy the same energy state which is described by the same quantum numbers. Thus it gives the basis for building up the elements in the periodic table in that one electron can occupy the K shell of hydrogen and two electrons (with opposite spins) can occupy the K shell of helium, but in lithium (the K shell having been filled) the third electron must occupy the L shell, and so on.

Thus we have a whole "sea" of negative energy states corresponding to $E \leqslant -m_0 c^2$, but they are filled and are unobservable. If, however, energy

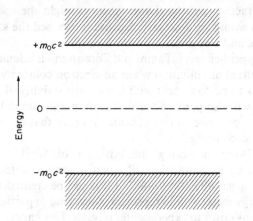

FIG. 3-14   Energy levels of relativistic electrons.

greater than $2m_0c^2$ be imparted to such a negative-energy electron, it can make the quantum transition from a negative-energy state and appear as a negative electron with positive energy. Since, however, our arbitrary zeros of charge and energy are as designated in Fig. 3-14, the "hole" that is left in the sea of negative-energy states appears as a positive charge with positive energy. Thus in elementary electrostatics when we remove negative electrons from a body which is initially at ground or zero potential, it acquires a positive potential and a net positive charge. Dirac also pointed out that the wave functions representing negative-energy states were such that "an electron with negative energy moves in an external (electromagnetic) field as though it carries a positive charge."

Initially Dirac thought that the "holes" represented protons, even though he was fully conscious of the difficulty that arose from the great discrepancy in mass. This view was also held by J. R. Oppenheimer, who suggested, however, in 1931 that *all* rather than *nearly all* (as had been suggested by Dirac) negative energy states in the world were filled.

The possibility of the negative-energy electron being connected in some way with the proton or hydrogen nucleus was entertained by a number of people, including Hermann Weyl. Weyl referred to this possibility in a 1929 publication, but in 1931 he came to the conclusion that, as attractive as the Dirac theory might be, some drastic modification would be necessary because it ascribed the same mass to both protons and electrons.

In 1930 Oppenheimer, I. Tamm and Dirac himself calculated that the chance of mutual annihilation when an electron collides with a "hole" was too great to be consistent with the known stability of protons and electrons. It was also pointed out that such mutual annihilation of two particles must give rise to *two* photons in order that both energy and momentum be conserved.

In 1931 Dirac, accepting the criticism of Weyl together with Oppenheimer's suggestion that all negative-energy states were filled, went on to suggest that the "hole," if it could be created in the sea of negative-energy states, would represent a new kind of particle which was at that time unknown to experimental physics. This particle would have the same mass as the electron and an equal but opposite charge. He called it an antielectron and went on to state that antiprotons could likewise be presumed to be possible. He suggested that an encounter between two γ rays, each of energy of at least 0.51 MeV (corresponding to the self-energy m_0c^2 given in Eq. 3-29), could lead to the simultaneous creation of an electron and an antielectron.

Two years later, in 1933, Dirac's brilliant prediction received confirmation by the discovery, in cloud-chamber photographs of cosmic radiation, of a positive electron by C. D. Anderson and of an electron-positron pair by P. M. S. Blackett and G. P. S. Occhialini.

Using the Einstein energy-mass equation, $E = mc^2$, the self-energy of an electron can easily be calculated to be 0.511 MeV. It is therefore possible for γ rays having energies greater than 1.022 MeV to raise a negative-energy electron to a positive-energy state and thereby create an electron-positron pair. This process is observed to occur with increasing efficiency as the γ-ray energy increases above 1.022 MeV. It must, however, take place within the Coulomb field of a nucleus (or electron, in the case of "triplet" production), so that the recoil of the nucleus (or electron) can allow both energy and momentum to be conserved. The process of pair formation is the reverse of that of bremsstrahlung production. In the former, a photon raises a negative-energy electron to a higher-energy state; in the latter, an electron passes to a lower-energy state and a photon is emitted. The decrease in photon intensity by pair

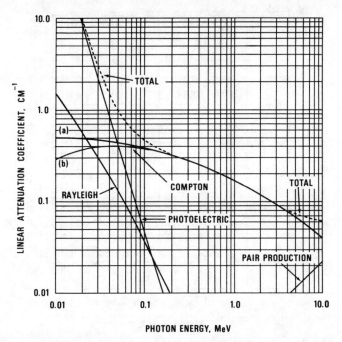

FIG 3-15   Linear attenuation coefficients as a function of photon energy for a *collimated, parallel beam* of monochromatic photons in aluminum. The mass attenuation coefficient can be derived by dividing the ordinates by $2.70 \, \text{g cm}^{-3}$, the density of aluminum. The "Compton" (a) and (b) curves are those pertaining to free and bound electrons, respectively. (These curves were plotted from data kindly supplied by Dr. John H. Hubbell.) (From NCRP, 1978).

production is thus given by equations very similar to Eqs. 3-8 and 3-9 (W. Heitler, *Quantum Theory of Radiation*).

We have now considered the three ways in which γ photons can transfer their energy to matter, namely, the photoelectric effect, the Compton effect, and the production of electron-positron pairs. Their relative contributions as a function of γ-ray energy are shown, for absorbers of both aluminum and lead, in Figs. 3-15 and 3-16. In these figures it will be noted that the term *attenuation* is used rather than *absorption*. An absorber both absorbs radiation and attenuates it, in proportions depending on the penetrability of radiation in matter. Thus

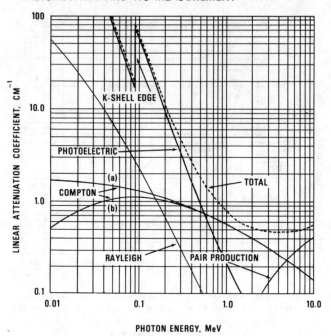

FIG. 3-16 Linear attenuation coefficients as a function of photon energy for a *collimated, parallel beam* of monochromatic photons in lead. The mass attenuation coefficient can be derived by dividing the ordinates by 11.34 g cm^{-3}, the density of lead. The total linear absorption coefficient is a minimum at about 4 MeV. The "Compton (a) and (b)" curves are those pertaining to free and bound electrons, respectively. (These curves were plotted from data kindly supplied by Dr. John H. Hubbell.) (From NCRP, 1978).

if we observe the amount of radiation, in a collimated parallel beam, that is transmitted through an absorber, α and β particles will be largely stopped, or absorbed out of the beam, whereas the more penetrating photons will be largely attenuated by scattering out of the beam.

As has been mentioned in the discussion of Compton scattering, photons can be coherently scattered without loss of energy. This is also known as *Rayleigh scattering.*

As with electrons, the *mass* coefficient for γ rays of low energy is almost independent of the material for low Z over a wide band of energies.

References

Allison, S. K. and Warshaw, S. D. (1953) Passage of heavy particles through matter, *Rev. Mod. Phys.*, *25*, 779.

Anderson, C. D. (1933) The positive electron, *Phys. Rev. 43*, 491.

Auger, P. (1926) Sur l'effet photoelectric compose, *J. de Phys.*, 6 [6], 205.

v. Baeyer, O. and Hahn, O. (1910) Magnetische Linienspektren von β-Strahlen, *Phys. Z.*, *11*, 488.

v. Baeyer, O., Hahn, O. and Meitner, L. (1911) Über die β-Strahlen des aktiven Niederschlags des Thoriums, *Phys. Z.*, *12*, 273.

v. Baeyer, O., Hahn, O. and Meitner, L. (1911) Nachweis von β-Strahlen bei Radium D, *Phys. Z.*, *12*, 378.

Bethe, H. A. and Ashkin, J. (1953) Passage of radiations through matter, *Experimental Nuclear Physics*, Segré, E., Ed. (John Wiley & Sons, Inc., New York).

Bethe, H. A. and Heitler, W. (1934) On the stopping of fast particles and on the creation of positive electrons, *Proc. Roy. Soc.*, A, *146*, 83.

Blackett, P. M. S. and Occhialini, G. P. S. (1933) Some photographs of the tracks of penetrating radiation, *Proc. Roy. Soc.*, A, *139*, 699.

Bloch, F. (1933) Zur Bremsung rasch bewegter Teilchen beim Durchgang durch Materie, *Ann. D. Physik, 16*, [5], 285.

Bohr, N. (1913) On the theory of the decrease of velocity of moving electrified particles on passing through matter, *Phil. Mag.*, *25* [6], 10.

Bohr, N. (1915) On the decrease of velocity of swiftly moving electrified particles in passing through matter, *Phil. Mag.*, *30* [6], 581.

Bragg, W. H. (1904) On the absorption of α rays, and on the classification of the α rays from radium, *Phil. Mag.*, *8* [6], 719.

Bragg, W. H. and Kleeman, R. (1904) On the ionization curves of radium, *Phil. Mag.*, *8* [6], 726.

Chadwick, J. (1914) Intensitätsverteilung im magnetischen Spektrum der β-Strahlen von Radium B + C, *Verh. der Deutsche Phys. Ges.*, *16*, 383.

Dirac, P. (1928) The quantum theory of the electron, *Proc. Roy. Soc.*, A, *117*, 618.

Dirac, P. (1930) On the annihilation of electrons and protons, *Proc. Camb. Phil. Soc.*, *26*, 361.

Dirac, P. (1930) A theory of electrons and protons, *Proc. Roy. Soc.*, A, *126*, 360.

Evans, R. D. (1955) *The Atomic Nucleus* (McGraw-Hill, New York).

Feather, N. (1938) Further possibilities for the absorption method of investigating the primary β-particles from radioactive substances, *Proc. Camb. Phil. Soc.*, *34*, 599.

Geiger, H. and Nuttall, J. M. (1911) The ranges of α particles from various radioactive substances and a relation between range and period of transformations, *Phil. Mag.*, 22 [6], 613.

Geiger, H. and Nuttall, J. M. (1912) The ranges of α particles from uranium, *Phil. Mag.*, *23* [6], 439.

Hubbell, J. H. (1969) *Photon Cross Sections, Attenuation Coefficients, and Energy Absorption Coefficients from 10 keV to 100 GeV*, Report NSRDS-NBS 29 (U.S. Government Printing Office, Washington).

Hubbell, J. H., Veigele, W. J., Briggs, E. A., Brown, R. T., Cromer, D. T. and Howerton, R. J. (1975) Atomic form factors, incoherent scattering functions, and photon scattering cross sections, *J. Phys. Chem. Ref. Data.*, *4*, 471.

Livingston, M. S. and Bethe, H. A. (1937) Nuclear physics: C. Nuclear dynamics, experimental, *Rev. Mod. Phys.*, *9*, 245.

Moseley, H. G. J. (1913) The high-frequency spectra of the elements I, *Phil. Mag.*, *26* [6], 1024.

Moseley, H. G. J. (1914) The high-frequency spectra of the elements II, *Phil. Mag.*, *27* [6], 703.

Moszkowski, S. A. (1965) Theory of multiple radiation, *Alpha-, Beta- and Gamma-Ray Spectroscopy*, 863, K. Siegbahn (Ed.), (North Holland Publishing Company, Amsterdam).

Mott, N. F. (1929) The exclusion principle and aperiodic systems, *Proc. Roy. Soc.*, A, *125*, 222.

Mott, N. F. (1930) The collision between two electrons, *Proc. Roy. Soc.*, A, *126*, 259.

NCRP (1978) *A Handbook of Radioactivity Measurements Procedures*, NCRP Report No. 58 (National Council on Radiation Protection and Measurements, Washington, D.C.)

Rutherford, E. (1914) The connexion between the β and γ ray spectra, *Phil. Mag.*, *28* [6], 306.

Rutherford, E. and Andrade, A. N. da C. (1914) The wave-length of the soft γ rays from radium B, *Phil. Mag.*, *27* [6], 854.

Rutherford, E. and Andrade, A. N. da C. (1914) The spectrum of the penetrating γ rays from radium B and radium C, *Phil. Mag.*, *28* [6], 263.

Rutherford, E., Robinson, H. and Rawlinson, W. F. (1914) Spectrum of the β rays excited by γ rays, *Phil. Mag.*, *28* [6], 281.

Sargent, B. W. (1933) The maximum energy of the β-rays from uranium X and other bodies, *Proc. Roy. Soc.*, A, *139*, 659.

Uhlenbeck, G. E. and Goudsmit, S. (1925) Ersetzung der Hypothese vom unmechanischen Zwang durch eine Forderung bezüglich des inner Verhaltens jedes einzelnen Elektrons, *Naturwissenschaften*, *13*, 953.

Uhlenbeck, G. E. and Goudsmit, S. (1926) Spinning electrons and the structure of spectra, *Nature*, *117*, 264.

Whitehouse, W. J. and Putman, J. L. (1953) *Radioactive isotopes: An introduction to their preparation, measurement and use* (The Clarendon Press, Oxford).

Widdowson, E. C. and Champion, F. C. (1938) Upper limits of continuous β-ray spectra, *Proc. Phys. Soc.*, *50*, 185.

4 The Neutrino and the Neutron

THE outstanding characteristic of β-ray spectra is that they are continuous in energy from zero to $E_{\beta_{max}}$, and this is true whether or not there is an associated γ ray resulting from the product nucleus being left in an excited state (i.e., a state in which the product nucleus is left with a surplus of energy above its ground state). This constituted one of the major problems of physics in the 1920's, because if a group of identical atoms of one element, identical, that is, in mass and energy level, decay to a group of atoms of an adjacent element, also all of the same mass and energy, why should not all the emitted electrons (β^-) or positrons (β^+) have the same energy?

One explanation that was given was that perhaps the β particles were emitted from the nucleus with the same energies but that some of this energy was dissipated by secondary processes such as extranuclear collisions. To test this hypothesis, C. D. Ellis and W. A. Wooster, in 1927, and also L. Meitner and W. Orthmann, in 1930, measured the rate of heat dissipation in radium E using microcalorimeters. If the energy of the β particle was indeed being dissipated in the same atom or in the surrounding atoms, then the microcalorimeter would also measure this energy as well as that from the absorption of the β particles in the calorimeter walls. However, in both experiments the microcalorimetrically measured rate of energy dissipation corresponded not with the maximum β-ray energy $E_{\beta_{max}}$, but with the average energy of the β particles. Where then did the rest of the energy go?

Bohr suggested that perhaps energy was not conserved in nuclear transitions, but in 1931 Pauli suggested that the disintegration energy was being carried away not only by the β particles, but also by particles having zero charge and small, if not zero, mass. Such particles would have little interaction with the surrounding matter and would thus

77

escape, undetected, from the microcalorimeter. These particles were first called *neutrons* by Pauli but were later referred to by Enrico Fermi as *neutrinos* (little neutral particles). In 1933 Ellis and Mott proposed that the difference in energy between radioactive and product nuclei was equal to $E_{\beta_{max}}$. Since the late 1940's the emission of the neutrino has been demonstrated indirectly by measuring the recoil of nuclei, after β emission or electron capture, for radionuclides that decay mainly to the ground state of the daughter nucleus, i.e., with no γ-ray emission. From the relationships between momenta and angle of the recoil nucleus and the β ray, the presence of a third momentum vector (that due to the neutrino) becomes necessary to conserve momentum. In the case of electron-capture, the momentum of the recoiling nucleus must be equal and opposite to that of the neutrino as these are the only two bodies involved. Since the decay is to specific states of the daughter nucleus, the neutrinos must also be monoenergetic. Two electron-capturing nuclides that have been used in such experiments are beryllium-7 and argon-37. Also, in 1956, C. L. Cowan and F. Reines successfully used the tremendous flux of neutrinos from a nuclear reactor to induce inverse β decay by bombarding hydrogen nuclei with neutrinos* and converting them into neutrons and positive electrons. As will also be discussed later, it is now clear that the maximum β-ray energy, $E_{\beta_{max}}$, corresponds to the energy available, as calculated by means of the Einstein relation, $E = mc^2$, from the difference in masses of the radioactive and product nuclei. The balance of energy released in the emission of a β particle having energy less than $E_{\beta_{max}}$ must therefore be emitted in some other form; and in a form that has little or no interaction with matter (such as that constituting the walls of a microcalorimeter).

Let us now turn briefly to the discovery of the neutron, a neutral particle that had been predicted in 1920 on the supposition that the charges of a proton and electron could neutralize to give a particle which would have approximately the same mass as the proton. Rutherford referred to this possible particle in the Bakerian lecture which he delivered to the Royal Society in London in 1920.

---

* For the same reason that the positron is the *antiparticle* of the electron the neutrino is believed to have an antiparticle known as the antineutrino. Generically we often use the term neutrinos to include both, but more strictly the anti-neutrino ($\bar{\nu}$) is associated with β^- and the neutrino (ν) with β^+ decay.

In 1930 W. Bothe and H. Becker observed that certain of the lighter elements, such as lithium and beryllium, emit a very penetrating radiation when they are bombarded by α particles from the natural radioactive elements. The absorption in lead of this penetrating radiation was found to be such that, *if* the radiation were electromagnetic, it would correspond to γ rays of about 10 MeV.

In 1932 F. Joliot and his wife, I. Curie, known first as the Curie-Joliots and latterly as the Joliot-Curies (and even once by a tongue-tied lecturer as the Jolie-Curiots), found that this strange new penetrating radiation, when incident upon hydrogenous material such as paraffin, possesses the property of being able to eject from it very energetic protons. Moreover, the energies of these protons are such as to make it necessary to assume that the incident radiation, if electromagnetic in nature, must consist of γ rays having energies of 50 MeV (if the ejected protons are assumed to be in the nature of Compton recoils).

There was thus a most serious discrepancy between the results of Bothe and Becker, giving "γ rays" of 10 MeV, and those of the Joliot-Curies, with a "γ-ray" energy of 50 MeV.

Also in 1932, H. C. Webster investigated the radiation from a number of the lighter elements, under α-particle bombardment, and found that the radiation, if electromagnetic, varied in energy from 0.5 MeV for sodium to 8 MeV for boron.

J. Chadwick also investigated these radiations and concluded that they were neutrons. On February 17, 1932, Chadwick sent a letter to *Nature* on the "Possible Existence of a Neutron." His paper on "The Existence of a Neutron" (a more confident title!) appeared in the June, 1932, *Proceedings of the Royal Society.*

Chadwick investigated the radiation resulting from the bombardment of beryllium with the α particles from polonium-210 (from radium D + E + F which he notes was "generously presented by Dr. C. F. Burnam and Dr. F. West, of the Kelly Hospital, Baltimore") using an ionization chamber connected to an amplifier and oscillograph, as illustrated in Fig. 4-1. He obtained about 4 deflections in the oscillograph per minute, and this rate was essentially unchanged by the interposition of 2 cm of lead. On interposing, instead of the lead, 2 cm of paraffin wax the number of deflections per minute increased markedly, and by carrying out an absorption experiment with aluminum the

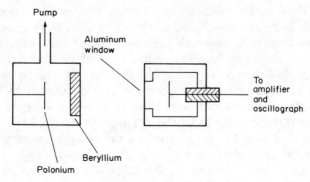

FIG. 4-1   Schematic drawing of source mount and ionization chamber used by Chadwick
in his discovery of the neutron.

ionizing radiation was found to consist of recoil protons. From the
known range-velocity curve for protons, these were deduced to have a
maximum recoil velocity of $3.3 \times 10^9 \, \text{cm s}^{-1}$, corresponding to an
energy of about 5.7 MeV. Chadwick also investigated the recoils from
other elements that were exposed to the radiation from the beryllium
and filled the ionization chamber with various gases, including nitrogen.
Knowing the proton recoil energy, and assuming them to be Compton
recoils, required that the incident γ rays should have energies of 55 MeV,
but γ rays of such an energy should give Compton recoils of nitrogen
atoms of some three times the energy actually observed by Chadwick. He
was therefore led to the conclusion that the radiation which was incident
upon the hydrogen and nitrogen must consist of neutral particles of mass
very nearly equal to that of the proton, which assumption, he pointed
out, resolved all the difficulties in the collision and recoil data. This
particle, he concluded, may be supposed "to consist of a proton and an
electron in close combination, the 'neutron' discussed by Rutherford in
his Bakerian lecture of 1920."

   In the paper directly following that of Chadwick, N. Feather describes
experiments in which he recorded some 100 cloud-chamber tracks of
recoil nitrogen atoms. (The uncharged neutron did not ionize the gas
molecules along its path, and so its track did not appear in the cloud-
chamber photographs.) The maximum range of these recoil atoms was

3.5 mm in air at STP, corresponding to a recoil velocity of $4.7 \times 10^8 \, \text{cm s}^{-1}$.

We thus have the maximum recoil velocities, u_p and u_n, of the hydrogen and nitrogen atoms equal to 3.3×10^9 and $4.7 \times 10^8 \, \text{cm s}^{-1}$, respectively. Moreover, we know that these maximum velocities must arise from head-on collisions. If M and V be the mass and velocity of the neutron, then for head-on collisions, we have

$$u_p = \frac{2M}{M+1} V, \tag{4-1}$$

and

$$u_n = \frac{2M}{M+14} V; \tag{4-2}$$

whence

$$\frac{u_p}{u_n} = \frac{M+14}{M+1} = \frac{3.3 \times 10^9}{4.7 \times 10^8} \tag{4-3}$$

From Eq. 4-3 M is readily calculated to be equal to 1.16 atomic mass units, when the proton and nitrogen atom are respectively 1 and 14 atomic mass units.

This was quite an approximate result, but by substituting powdered boron deposited on a graphite plate for the beryllium target shown in Fig. 4-1 Chadwick obtained a more accurate value for M. A sheet of paraffin wax was then interposed between the target and ionization chamber, and the maximum range and velocity of the recoil protons were determined, as before, by their absorption in aluminum. The velocity was found to be $2.5 \times 10^9 \, \text{cm s}^{-1}$, and (by substituting $M \sim 1$ in Eq. 4-1) we see that this must also be the maximum velocity of the neutrons. Chadwick assumed that the nuclear reaction taking place was

$$^{11}_{5}\text{B} + {}^{4}_{2}\text{He} \rightarrow {}^{14}_{7}\text{N} + {}^{1}_{0}\text{n}. \tag{4-4*}$$

The masses of ^{11}B and ^{14}N were known from the mass-spectrometer measurements of F. W. Aston. The velocity of the polonium-210 α particle was $1.59 \times 10^9 \, \text{cm s}^{-1}$, and the maximum neutron velocity was $2.5 \times 10^9 \, \text{cm s}^{-1}$. Assuming conservation of momentum in the collision,

* It is conventional to denote the whole-number atomic mass as a superscript and the atomic number as a subscript before the chemical symbol.

the velocity of the recoiling ^{14}N atom was calculated by Chadwick, who then wrote the following energy equation for the reaction:

Mass of ^{11}B + mass of ^{4}He + kinetic energy of ^{4}He = mass of ^{14}N + mass of ^{1}n + kinetic energy of ^{14}N + kinetic energy of ^{1}n.

$$(4\text{-}5)$$

From Aston's results the masses in atomic mass units (amu) were 11.00825 for ^{11}B, 4.00106 for ^{4}He, and 14.0042 for ^{14}N. The kinetic energies in Eq. 4-5 must be converted to mass units by the Einstein relation, $E = mc^2$ (from which it is readily calculated that an energy of 931 MeV is equivalent to 1 amu*). The kinetic energies in atomic mass units were 0.00565 for the α particle, 0.0035 for the neutron, and 0.00061 for the nitrogen atom. Substituting these six quantities in Eq. 4-5 gave the mass of the neutron equal to 1.0067 atomic mass units. Nowadays the most accurate results for the respective masses of the neutron and proton on the carbon-12 scale are taken as 1.008665 and 1.007277.

Chadwick was unable to deduce the mass of the neutron from his results for beryllium because the nuclear reaction which he presumed to occur in this case was

$$^9_4Be + {}^4_2He \rightarrow {}^{12}_6C + {}^1_0n, \qquad (4\text{-}6)$$

and the mass of the beryllium atom was not then known.

The interactions of neutrons with matter are not as pronounced as, say, those of protons or other charged particles in motion. The neutron has little effect on the extranuclear electrons and is unaffected by the electric field of the nucleus. Only when it approaches to within distances equal to the order of 10^{-13} cm does it experience the strong attractive nuclear force and interact with it. Figure 4-2 shows schematically a cross section of atoms of an absorber, with the path AB of a particle traversing it. If AB represented a proton or α particle, electrons would be detached

---

* The value of 1 amu is 1.6606×10^{-27} kg, and c is equal to 2.998×10^8 m s^{-1}. The product mc^2 is therefore equal to 1.492×10^{-10} J. The charge on the electron is 1.602×10^{-19} C, and 1 V equals 1 J/C. Thus 1 eV is equivalent to 1.602×10^{-19} J and 1 MeV is equivalent to 1.602×10^{-13} J. But 1 amu is equivalent to 1.492×10^{-10} J, so that 1 amu is equivalent to $(1.492/1.602) \times 10^3$ MeV; i.e., 1 amu $\sim$ 931 MeV. Likewise 1 electron mass ~ 0.511 MeV.

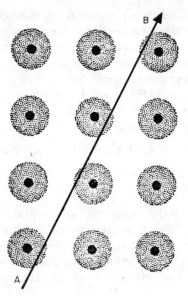

FIG. 4-2   Schematic of the cross section of a solid lattice with atoms consisting of nuclei surrounded by electron clouds. In a gas the atoms, or molecules, would have random distributions.

from the atoms and a trail of ion pairs would be left behind. A neutron, however, passes along the same trajectory with very little interaction unless it happens to collide with a nucleus. In this event the recoiling nucleus gives rise to dense ionization.

Until 1932 it had been assumed that nuclei were composed of protons and electrons, but this assumption had led to a number of difficulties. With the discovery of the neutron, however, these difficulties were resolved by assuming, as did D. D. Iwanenko and Heisenberg, that nuclei are composed of protons and neutrons, or *nucleons* as they are now called.

In 1934, Fermi formulated a theory of β decay based on the assumption that β decay is the result of a neutron changing to a proton *in the nucleus*, i.e.,

$$\ _0^1n \rightarrow \ _1^1H + \beta^- + \bar{\nu}. \tag{4-7}$$

The β-ray spectrum such as is shown in Fig. 3-7 shows the distribution of numbers of β particles as a function of their energy. It thus also shows the *probability* that a β particle of a given energy will be emitted as well as the distribution of energy between the β particles and the antineutrinos, since the *total* energy released in each β event is $E_{\beta_{max}}$. Fermi found that, if W be the total energy of the β particle (including its rest mass) in units of m_0c^2, then the probability $P\,dW$ of the emission in unit time of a β particle with energy between W and $W + dW$ was

$$P\,dW = kf(Z, W)pW(W_{\beta_{max}} - W)^2\,dW, \qquad (4\text{-}8)$$

where k is a constant and p is the momentum of the β particle in units of m_0c $(p = \sqrt{W^2 - 1}\,)$.

In Eq. 4-8, $W_{\beta_{max}}$ is the energy released by the decaying nucleus, and $(W_{\beta_{max}} - W)$ is therefore the energy carried away by the neutrino, both in units of m_0c^2. The function $f(Z, W)$ involves the effect of the Coulomb field of the nucleus on the emitted β particle. The value of $f(Z, W)$ is equal to unity for $Z = 0$, and is not much different from unity for $Z < 20$, and a plot of

$$P\,dW \propto pW(W_{\beta_{max}} - W)^2\,dW \qquad (4\text{-}9)$$

is of the form shown in Fig. 4-3, which approximates the shape of a great many experimentally observed β spectra when due allowance is made for

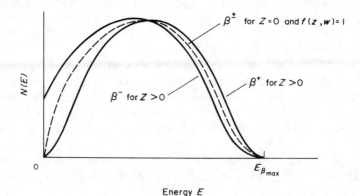

FIG. 4-3   Frequency distribution of β rays as a function of energy (the β-ray spectrum), as given by the Fermi theory of β decay.

the effect of the Coulomb field of the nucleus. This effect is to slow down β^- particles and therefore to give an excess of them at lower energies. Fermi's theory can also be applied to positron decay. In this case the β^+ particles are accelerated away from the nucleus and there is a deficit of them at the lower energies, as illustrated in Fig. 4-3.

References

Bothe, W. and Becker, H. (1930) Künstliche Erregung von Kern-γ-Strahlen, *Z. Phys.*, *66*, 289.

Bothe, W. and Becker, H. (1930) Aufbau von Atomkernen, *Naturwissenschaften*, *19*, 753 (1931).

Chadwick, J. (1932) Possible existence of a neutron, *Nature 129*, 312.

Chadwick, J. (1932) The existence of the neutron, *Proc. Roy. Soc.*, A, *136*, 692.

Curie, Irène and Joliot, F. (1932) Emission de protons de grande vitesse par les substances hydrogenées sous l'influence des rayons γ très pénétrants, *Comptes Rendus, Académie des Sciences, 194*, 273.

Cowan, C. L. and Reines, F. (1953) Detection of the free neutrino, *Phys. Rev. 92*, 830.

Ellis, C. D. and Mott, N. F. (1933) Energy relations in the β-ray type of radioactive disintegration, *Proc. Roy. Soc.*, A, *141*, 502.

Ellis, C. D. and Wooster, W. A. (1927) The average energy of disintegration of radium E, *Proc. Roy. Soc.*, A, *117*, 109.

Feather, N. (1932) The collisions of neutrons with nitrogen nuclei, *Proc. Roy. Soc.*, A, *136*, 709.

Fermi, E. (1934) Versuch einer Theorie der β-Strahlen. I., *Zeit. f. Phys.*, *88*, 161.

Meitner, L. and Orthmann, W. (1930) Über eine absolute Bestimmung der Energie der primären β-Strahlen von Radium E, *Z. Phys.*, *60*, 143.

Pauli, W. (1933) *Rapports du Septième Conseil de Physique Solvay*, Brussels 1933 (Gauthier-Villars et Cie, Paris, 1934).

Webster, H. C. (1932) The artificial production of nuclear γ-radiation, *Proc. Roy. Soc.*, A, *136*, 428.

5 The Energetics
of Nuclear Change

IN the last chapter, and earlier, we have jumped a little ahead of our historical account of the development of radioactivity, by talking about artificially produced radionuclides, the products of nuclear reactions.

The first example of an artificially produced nuclear reaction, or disintegration, was that observed by Rutherford in 1919 when he used a source of radium C + C' (with α-particle energies of 7.9 MeV) to bombard nitrogen. As a result, long-range particles were produced corresponding to protons (which were also identified by magnetic deflection) of a range of 28 cm in air at STP. Care was taken to purify the nitrogen of any trace of hydrogen in order to remove the possibility of any hydrogen recoils. Subsequently, Rutherford and Chadwick found that the long-range protons had a range of 40 cm in air at STP, compared with 29 cm for the maximum possible range for any recoil protons. These observations are represented by the nuclear reaction

$$^{14}_{7}\text{N} + {}^{4}_{2}\text{He} \rightarrow {}^{17}_{8}\text{O} + {}^{1}_{1}\text{H}. \tag{5-1}$$

The next advance in this field came in 1932 when J. D. Cockcroft and E. T. S. Walton, in Rutherford's laboratory, bombarded lithium with 300-keV protons produced in a discharge tube and observed, by means of a cloud chamber, that α particles were produced according to the reaction

$$^{7}_{3}\text{Li} + {}^{1}_{1}\text{H} \rightarrow 2{}^{4}_{2}\text{He}. \tag{5-2}$$

The α particles appeared as pairs in opposite directions, thereby conserving energy and momentum, with equal ranges of about 8.3 cm in air at STP. These ranges correspond to an initial kinetic energy of the α particle of about 8.55 MeV.

The balance of mass and energy in this nuclear reaction can be expressed by means of the following dimensionally somewhat hybrid, but unambiguous, reaction:

$$_3^7\text{Li} + {}_1^1\text{H} + 300\,\text{keV} \rightarrow 2_2^4\text{He} + 17.1\,\text{MeV}. \qquad (5\text{-}3)$$

From the Einstein mass-energy relation, $E = mc^2$, we have already calculated that 1 atomic-mass unit (amu) is equivalent to 931 MeV, whence

17.1 MeV is equivalent to 0.0184 amu,

300 keV is equivalent to 0.0003 amu.

In the nuclear reaction shown in Eq. 5-3 the nuclei are stripped of extranuclear electrons. The masses of $_1^1\text{H}$, $_2^4\text{He}$, and $_3^7\text{Li}$ are thus atomic masses less the mass of one, two, and three electrons, respectively. On the carbon-12 scale these masses were:

$_1^1\text{H}^+ \quad = 1.00783 - 0.00055 = 1.0073\,\text{amu};$

$_2^4\text{H}^{++} \quad = 4.00260 - 0.00110 = 4.0015\,\text{amu};$

$_3^7\text{Li}^{+++} = 7.0161 - 0.00165 = 7.0144\,\text{amu}.$

(The electron rest mass = 0.00055 amu.)

The mass balance of Eq. 5-3 is thus:

Left-hand side: $7.0144 + 1.0073 + 0.0003 = 8.0220\,\text{amu};$
Right-hand side: $8.0030 + 0.0184 \qquad\qquad = 8.0214\,\text{amu}.$

This also affords experimental proof of the applicability of the Einstein relation, $E = mc^2$, to nuclear reactions, an assumption previously made by Chadwick in determining the mass of the neutron.

With the discovery of the neutron in 1932 it was immediately observed by Feather in his cloud-chamber experiments with nitrogen recoils that some of the recoils were not from elastic collisions, but that two tracks resulted, corresponding to the reaction

$$_7^{14}\text{N} + {}_0^1\text{n} \rightarrow {}_5^{11}\text{B} + {}_2^4\text{He}. \qquad (5\text{-}4)$$

It was soon found that the uncharged neutron which was unaffected by the Coulomb field of the nucleus was very effective in producing nuclear transformations.

In 1934 F. Joliot and I. Joliot-Curie, when studying the effects of the bombardment of light nuclei with the α particles from polonium-210, observed that both positrons and neutrons were emitted. This was in itself not surprising as it appeared to be closely similar to the reaction discovered by Rutherford in 1919 (Eq. 5-1), as the positron and the neutron together could be considered to be equivalent to the proton observed by Rutherford.

However, the surprising fact was soon noted by the Joliot-Curies that when the α-particle source was removed the emission of neutrons ceased *but* the positrons continued to be emitted, and what is more, *their activity decayed exponentially*!

Thus was artificial, or induced, radioactivity discovered. The elements which, in particular, were found by the Joliot-Curies to give rise to this phenomenon on bombardment with α particles were boron, magnesium, and aluminum, with the following reactions:

$$^{10}_{5}B + {}^{4}_{2}He \rightarrow {}^{13}_{7}N + {}^{1}_{0}n, \tag{5-5}$$

$$^{24}_{12}Mg + {}^{4}_{2}He \rightarrow {}^{27}_{14}Si + {}^{1}_{0}n, \tag{5-6}$$

$$^{27}_{13}Al + {}^{4}_{2}He \rightarrow {}^{30}_{15}P + {}^{1}_{0}n. \tag{5-7}$$

The nuclei ^{13}N, ^{27}Si, and ^{30}P are all positron-active and decay according to the decay schemes shown in Fig. 5-1. Since the Joliot-Curies' discovery a great many positron and electron emitters have been found, and a few radionuclides are now known to decay by both β^+ *and* β^- emission. One such radionuclide is copper-64.

In 1936 J. J. Livingood artificially produced, for the first time, a naturally-occurring radioactive element (radium E, or bismuth-210) from a stable nuclide by deuteron bombardment of bismuth-209, by the (d, p)* reaction

$$^{209}_{83}Bi + {}^{2}_{1}H \rightarrow {}^{210}_{83}Bi + {}^{1}_{1}H. \tag{5-8}$$

---

* (d, p) represents "deuteron in and proton out."

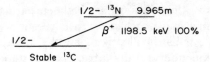

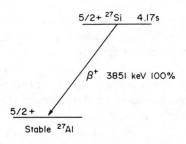

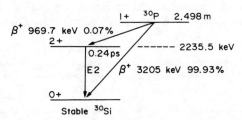

FIG. 5-1 Nuclear decay schemes for nitrogen-13, silicon-27, and phosphorus-30.

In 1935, Fermi, E. Amaldi, O. D'Agostino, B. Pontecorvo, F. Rasetti, and E. Segré discovered (and patented) the way to produce slow neutrons. In general, a neutron which suffers an elastic collision with a heavy nucleus does not lose much energy. (A tennis ball which bounces off a brick wall does so with very little loss of kinetic energy.) Fermi and his colleagues found, however, that when fast neutrons (as ordinarily produced by the nuclear reactions already mentioned) were allowed to pass through an hydrogenous material such as water or paraffin wax the neutrons were slowed down and became considerably more effective in producing nuclear reactions. When neutrons collide elastically with hydrogen atoms, the collisions are between particles of nearly equal mass, and on an average, each collision halves the energy of the neutron. (For example, a billiard ball striking another billiard ball head-on can transfer all its kinetic energy to the ball that was stationary. On the other hand, a glancing collision will transfer very little, and the average transfer is around one half of the kinetic energy per collision.)

Thermal neutrons, which are neutrons in thermal equilibrium with the paraffin wax, have a normal Maxwellian distribution of energies as given by the kinetic theory. P. B. Moon and J. R. Tillman found that paraffin wax cooled in liquid nitrogen made the neutrons even more effective in producing certain radionuclides. They subsequently showed, with G. P. Thomson and E. C. Wynn-Williams, that the Maxwellian distribution of velocities shifted with the reduction in temperature to a lower value of the average energy. They concluded therefore that such neutrons were in thermal equilibrium with the paraffin wax, and called them thermal neutrons. Their average energy is $3kT/2$ where k is Boltzmann's constant and T is the absolute temperature.

Fast, slow and thermal neutrons lie in the approximate energy ranges shown in Table 5-1, with corresponding values of the mean free path.

TABLE 5-1   *Approximate Energy Ranges of Fast, Slow, and Thermal Neutrons*

Type of Neutron	Energy	Mean Free Path in Paraffin Wax
Fast	1 to 2 MeV	$\sim 3\,\mathrm{cm}$
Slow	1 eV	$\sim 0.6\,\mathrm{cm}$
Thermal	0.025 eV	$\sim 0.4\,\mathrm{cm}$

There are many examples of slow-neutron capture, usually with the emission of a γ ray to get rid of excess energy in the product nucleus. In line with our convention with deuteron reactions (i.e., (d, n) and (d, p)) such capture of a neutron with an emission of a γ ray is designated (n, γ). One such typical reaction is

$$\mathrm{{}^1_1H} + \mathrm{{}^1_0n} \rightarrow \mathrm{{}^2_1H} + h\nu, \tag{5-9}$$

which is the reverse reaction to the photo-disintegration of the deuteron, which was utilized by Chadwick and M. Goldhaber to measure the mass of the neutron. The energy of the slow neutron is only of the order of an electron volt so that the energy available for the γ ray ($h\nu$) arises only from the binding energy of the deuteron (i.e., the energy given out when a neutron and a proton form a nucleus). This energy is of the order of 2 MeV.

Neutrons are not easily detected by the general method of electrical counting because they do not ionize. They must be detected by their secondary effects. Thus an electrical counter or ionization chamber containing helium-3 or boron trifluoride will detect slow neutrons satisfactorily by virtue of the reactions

$$\mathrm{{}^{3}_{2}He + {}^{1}_{0}n \rightarrow {}^{3}_{1}H + {}^{1}_{1}H,} \tag{5-10}$$

$$\mathrm{{}^{10}_{5}B + {}^{1}_{0}n \rightarrow {}^{7}_{3}Li + {}^{4}_{2}He.} \tag{5-11}$$

In each case the recoiling nuclei ionize the gas filling of the counter or ion chamber. The helium-3 reaction has an exceptionally high probability of taking place.

Another method of detecting slow neutrons is by their reaction with manganese as a salt in solution or with a foil of indium or gold, i.e.,

$$\mathrm{{}^{55}_{25}Mn + {}^{1}_{0}n \rightarrow {}^{56}_{25}Mn + h\nu,} \tag{5-12}$$

$$\mathrm{{}^{115}_{49}In + {}^{1}_{0}n \rightarrow {}^{116m}_{49}In + h\nu,} \tag{5-13}$$

$$\mathrm{{}^{197}_{79}Au + {}^{1}_{0}n \rightarrow {}^{198}_{79}Au + h\nu.} \tag{5-14}$$

The indium-116m is an *isomeric state* of indium-116, that is to say, a state which exists for a finite time above the ground-level of the same nuclide. Most isomeric levels decay to the ground level by simple emission of a γ ray. However, indium-116m, manganese-56, and gold-198 all decay by β^- emission which can be detected by an electrical counter or ionization chamber:

$$\mathrm{{}^{56}_{25}Mn \xrightarrow{t_{1/2}\,=\,2.6\,hours} {}^{56}_{26}Fe + \beta^- + \overline{\nu};} \tag{5-15}$$

$$\mathrm{{}^{116m}_{49}In \xrightarrow{t_{1/2}\,=\,54\,min} {}^{116}_{50}Sn + \beta^- + \overline{\nu};} \tag{5-16}$$

$$\mathrm{{}^{198}_{79}Au \xrightarrow{t_{1/2}\,=\,2.7\,days} {}^{198}_{80}Hg + \beta^- + \overline{\nu}.} \tag{5-17}$$

Nuclear reactions are an extremely sensitive way of checking nuclear masses because a large and easily measured quantity of energy is equivalent to quite a small difference in mass (i.e., $0.1\,\text{MeV} \cong 0.0001\,\text{amu}$). As a result of investigations of such reactions it was suggested in 1935 both by H. A. Bethe and by M. L. E. Oliphant, A. R. Kempton, and Lord Rutherford that some of the mass-spectrometric masses of the lighter elements were in need of correction. Prior to 1935 the mass of the helium-4 atom had been taken to be 4.00216 amu (on the physical atomic-weight scale for which $^{16}\text{O} = 16\,\text{amu}$). It was important that this value should be correctly known as it was used as a substandard in mass-spectrometric measurements at low A. As a result of experimental investigations of nuclear reactions involving beryllium and boron by Oliphant, Kempton, and Rutherford the mass of helium-4 was redetermined mass-spectrometrically and found to be 4.00387 (for $^{16}\text{O} = 16\,\text{amu}$). By proton bombardment of beryllium and boron and by deuteron bombardment of boron the following reactions were observed:

$$^{9}_{4}\text{Be} + {}^{1}_{1}\text{H} \rightarrow {}^{8}_{4}\text{Be} + {}^{2}_{1}\text{H} + Q(= 0.46\,\text{MeV}), \qquad (5\text{-}18)$$

$$^{11}_{5}\text{B} + {}^{1}_{1}\text{H} \rightarrow {}^{8}_{4}\text{Be} + {}^{4}_{2}\text{He} + Q(= 8.70\,\text{MeV}) \qquad (5\text{-}19)$$

$$^{11}_{5}\text{B} + {}^{2}_{1}\text{H} \rightarrow {}^{9}_{4}\text{Be} + {}^{4}_{2}\text{He} + Q(= 8.13\,\text{MeV}). \qquad (5\text{-}20)$$

The values of Q are derived from the observed product energies *less* the energy of the bombarding proton or deuteron. If we now add reaction 5-18 to reaction 5-20, so to speak, algebraically, we get reaction 5-19. Therefore the sum of the values of the net Q for reactions 5-18 and 5-20 (namely, $0.46 + 8.13 = 8.59\,\text{MeV}$) should cross-check with the net value of Q for reaction 5-19. It does so to the order of 0.0001 amu. Further, by putting the then accepted values of the atomic masses individually into reactions 5-18, 5-19, and 5-20, a check can be made for the energy released in each. A number of other such triads can be formed with reactions involving additional lighter elements such as lithium-6 and lithium-7.

A reaction such as 5-9 is exoergic, or exothermic in the same way as chemical reactions can be exothermic and can proceed with the release of energy. The reverse reaction can be made to occur by supplying external energy in the form of γ radiation. The *decrease* in mass (of ^{2}H compared with $^{1}\text{H} + {}^{1}\text{n}$) in reaction 5-9 is directly converted into some 2 MeV of energy in the form of γ radiation.

We can apply precisely similar methods to derive the binding energy of the α particle. On the $^{12}C = 12$ scale, the proton, neutron, and helium-4 masses are respectively 1.007825, 1.008665 and 4.002604 amu. If B is the binding energy, then for the α particle

$$B = (2 \times 1.007825 + 2 \times 1.008665 - 4.002604) \times 931 \text{ MeV}$$

$$= (4.032980 - 4.002604) \times 931 \text{ MeV}$$

$$= 28.3 \text{ MeV}.$$

As the binding energy of the deuteron is only about 2 MeV, we see that the α particle is a much more stable unit. (In the above calculation for B we have taken *atomic* rather than *nuclear* masses because the electrons cancel anyhow. One must be careful to be consistent in this respect.)

The binding energy is thus the amount of energy required to take all the nucleons in a nucleus apart or, as we usually say, to remove them to an infinite distance from each other. Thus the binding energy B is given by

$$B = (NM_n + ZM_p - M)c^2, \qquad (5\text{-}21)$$

where N is the number of neutrons of mass M_n, Z is the number of protons of mass M_p, M is the mass of the nucleus in question, and c is the velocity of light. It is convenient to define the *binding energy per nucleon* or the *binding fraction* $f = B/A$, where A is the mass number. The binding energy per nucleon or binding fraction is thus equal to 1.1 MeV for the deuteron, 7.1 MeV for the α particle, and 8.0 MeV for oxygen-16 (i.e., 128 MeV/16). For oxygen-16

$$B = (8 \times 1.007825 + 8 \times 1.008665 - 15.99491) \times 931 \text{ MeV}$$

$$= 127.6 \text{ MeV}$$

The binding energy per nucleon is plotted as a function of atomic-mass number (A) in Fig. 5-2. The maximum values correspond to the more stable nuclei, but the values from $A = 20$ to $A = 190$ are above 8 MeV while between $A = 30$ and $A = 100$ the binding fraction is fairly constant between 8.7 and 8.8 MeV. The decrease in binding fraction with increasing atomic mass, above $A = 60$, occurs because the increased Coulomb repulsion makes the nuclei less stable. The rapid drop in

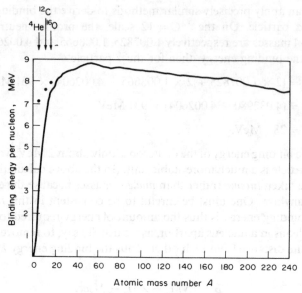

FIG. 5-2   Binding energy per nucleon in MeV as a function of the atomic mass number A.

binding fraction below about $A = 8$ (apart from helium-4) occurs because each nucleon is attracted by only a few other nucleons; the separation between them therefore increases and stability decreases. Thus the binding fraction of the deuteron is only 1.1 MeV and that of the proton is, of course, zero. The maximum of the binding fraction curve is in the region of $A = 55$ to $A = 65$. As we have already noted, this corresponds to the nuclei of maximum stability. It has been pointed out that this may have some bearing on evidence that the core of the earth may be mainly of iron and nickel for which A varies from 54 to 58 and from 58 to 64, respectively.

Because helium-4 (which falls slightly off the binding fraction curve) has a binding fraction of 7.1 MeV and because this is energy *released* per nucleon as the two protons and two neutrons come in from infinity (releasing 28.3 MeV per α particle), it is clear that if we could *fuse* protons and neutrons together to form α particles, then we would release large amounts of energy. Because hydrogen is one of the most abundant elements in the universe, such a fusion process would supply enormous amounts of energy. It has long been thought that this is the process which supplies stellar energy, in particular the energy from our own sun. It was difficult to understand, however, how such a fusion process could explain the large quantities of energy emitted by the stars when the probability of two protons and two neutrons colliding at the same time was so small. One could not imagine the process being one of building two deuterons which in turn would form an α particle because deuterons, with their low binding energies, would probably dissociate before they had a chance to collide.

In 1939, Bethe therefore proposed the "carbon cycle" for the production of stellar energy, in which the carbon is merely a vehicle or catalyst for the formation of the α particles. This process, which does give correct results within the present limits of uncertainty, is as follows:

$$^{12}C + {}^{1}H \rightarrow {}^{13}N + \gamma;$$

$$^{13}N \rightarrow {}^{13}C + \beta^{+} + \nu;$$

$$^{13}C + {}^{1}H \rightarrow {}^{14}N + \gamma;$$

$$^{14}N + {}^{1}H \rightarrow {}^{15}O + \gamma;$$

$$^{15}O \rightarrow {}^{15}N + \beta^{+} + \nu;$$

$$^{15}N + {}^{1}H \rightarrow {}^{12}C + {}^{4}He.$$

Thus we finish with a carbon-12 atom to replace the one we started with *and* an α particle. The net energy released in the process appears in the form of γ rays and in the kinetic energies of the particles emitted. The half lives of the two artificially radioactive elements produced, ^{13}N and ^{15}O, are respectively about 10 minutes and 2 minutes.

Concepts which are quite analogous to *binding energy* and *binding fraction* are those of *mass defect* and *packing fraction*. If Δ be the mass defect, then

$$\text{mass defect } \Delta = M - A,$$

and

$$\text{packing fraction} = (M - A)/A,$$

where M is the mass of the nucleus in amu and A is the atomic-mass number. These quantities have one disadvantage, however, in that they are not independent of the atomic-mass scale ($^{12}C = 12$ or $^{16}O = 16$). Nor is the mass defect so fundamental a quantity as the binding energy since, in the latter, instead of finding the difference between M and A (A being a whole number), we calculate the difference (Eq. 5-21) between M and $(NM_n + ZM_n)$. In other words $B \neq 931\Delta$ in MeV. Nevertheless, the packing-fraction curve is quite similar to the binding-fraction curve, except that it is "upside-down" and has a minimum in the region of atomic-mass number in which the binding-fraction curve is a maximum. The minimum in the packing-fraction curve therefore corresponds to maximum stability. While the reader may already be acquainted with the packing-fraction curve and should realize its connection with the binding-fraction (binding-energy-per-nucleon) curve, it will not be considered further here.

The last examples of mass-energy relations that we wish to consider, as such, are concerned with the conservation of mass-energy in α, β^-, and β^+ decay, including the process of electron capture. Let us, first of all, consider α decay.

Even though radioactivity involves a *nuclear* rearrangement, the mass-energy balances are based on the initial and final states of the parent and daughter *atoms*. Thus if a daughter nucleus is left in an excited state with excess energy that can be emitted as γ radiation, the excess energy can also be dissipated by an interaction in which this energy can be transferred to an extranuclear electron which is ejected as a so-called conversion electron. The process of electron capture is another example of a direct interaction between the nucleus and one of its extranuclear, or atomic, electrons.

In considering mass-energy balances it is important to remember that nuclear and atomic masses are linked by the equation

$$M_A^Z = m_A^Z + Zm_e - B_A^Z/c^2, \tag{5-22}$$

where M_A^Z and m_A^Z are the atomic and nuclear masses of a nuclide of atomic number Z and atomic-mass number A, m_e is the mass of an electron, and B_A^Z is the binding energy of the atom's Z electrons. We thus see that M_A^Z is less in mass than $m_A^Z + Zm_e$ by B_A^Z/c^2, which is *emitted* as the Z electrons come from infinity.

In order to calculate the energy release in α and β decay it is necessary to make certain basic assumptions about the initial and final states and also about the order of the times involved in such processes. We shall assume that the processes of extranuclear-electron rearrangement occur subsequent to, and thus independently of, the nuclear transitions.

We shall therefore assume that the radioactive parent consists of atoms in their *ground state* whereas the daughter atoms may be doubly, negatively, ionized in α decay, singly negatively ionized in β^+ decay, and singly positively ionized in β^- decay. The maximum amount of energy available, which is our only concern, is that for a transition to the *ground state* of any of these ions. In electron capture the decay is from Z to $Z - 1$, and the remaining number of extranuclear electrons is also $Z - 1$.

It is now possible to consider the energetics of the various modes of radioactive decay. In α decay

$$Z^A \rightarrow (Z - 2)^{A-2^{--}} + \alpha,$$

where A is the mass number. As no energy is supplied externally, the maximum amount of energy available for this transformation, Q_α, is, in mass units,

$$Q_\alpha = M_A^Z - (M_{A-4}^{Z-2})^{--} - m_\alpha,$$

and α decay can therefore occur if Q_α is positive, or if

$$M_A^Z > (M_{A-4}^{Z-2})^{--} + m_\alpha,$$

or

$$M_A^Z > M_{A-4}^{Z-2} + m_\alpha + 2m_e - I^{--}/c^2,$$

where

$$I^{--} = B_{A-4}^{Z-2} - (B_{A-4}^{Z-2})^{--}.$$

In general, the "mass" of double electron attachment, I^{--}/c^2, can be neglected compared with $M_{A-4}^{Z-2} + m_\alpha + 2m_e$, and α decay is possible if

$$M_A^Z > M_{A-4}^{Z-2} + m_\alpha + 2m_e. \tag{5-23}$$

In β^- decay

$$Z^A \rightarrow (Z + 1)^{A^+} + \beta^- + \bar{\nu}.$$

Assuming that the antineutrino has zero mass and that the transition is to the ground state of the singly ionized atom $(Z + 1)^A$, the maximum energy available for the transition (which energy will be shared by the antineutrino and the β^- particle, to total $E_{\beta_{max}}$, and in recoil of the nucleus) is

$$Q_{\beta^-} = M_A^Z - (M_A^{Z+1})^+ - m_e,$$

where m_e is the rest mass of the β^- particle. As no external energy is given to the system, β^- decay can occur only if Q_{β^-} is positive, or if

$$M_A^Z > (M_A^{Z+1})^+ + m_e$$

or

$$M_A^Z > M_A^{Z+1} + I^+/c^2,$$

where

$$I^+ = B_A^{Z+1} - (B_A^{Z+1})^+,$$

and

$$(M_A^{Z+1})^+ + m_e - I^+/c^2 = M_A^{Z+1}.$$

In general, the term I^+/c^2 can be neglected and β^- decay is possible if

$$M_A^Z > M_A^{Z+1}. \tag{5-24}$$

In the β^- decay of phosphorus-32 to sulphur-32 the atomic masses on the carbon-12 scale are respectively 31.973908 amu and 31.972072 amu, giving a *decrease* in mass on β^- decay of 0.001836 amu. This decrease corresponds to an energy release of 1.709 MeV, which is in good agreement with the experimentally observed value of $E_{\beta_{max}}$ for phosphorus-32 of about 1.710 MeV (corresponding to zero energy of the antineutrino). Potassium-40 is some 0.000807 amu heavier than calcium-40, and the former is found to decay to the latter by β^- emission with a half life of about 1.3×10^9 years.

In β^+ decay

$$Z^A \rightarrow (Z - 1)^{A^-} + \beta^+ + v,$$

and the maximum energy available is

$$Q_{\beta^+} = M_A^Z - (M_A^{Z-1})^- - m_e,$$

where m_e is the rest mass of β^+ particle. Again, as no external energy is supplied, the transition can only occur if

$$M_A^Z > (M_A^{Z-1})^- + m_e$$

or

$$M_A^Z > M_A^{Z-1} + 2m_e - I^-/c^2,$$

where

$$I^- = (B_A^{Z-1})^- - B_A^{Z-1}$$

and

$$M_A^{Z-1} + m_e - I^-/c^2 = (M_A^{Z-1})^-.$$

Again, the term I^-/c^2 can be neglected, and β^+ decay is possible if

$$M_A^Z > M_A^{Z-1} + 2m_e. \tag{5-25}$$

The difference of $2m_e$ between relations 5-24 and 5-25 is that in β^+ decay the system loses *two* electrons, a positron from the nucleus and an extranuclear electron that is superfluous to the $(Z - 1)$ atom. On the other hand β^- decay should *just* be able to occur if the β^- particle gets out of the nucleus with sufficient energy to supply the needed extranuclear electron for atom $(Z + 1)$, the antineutrino having zero energy. A detailed study of the above relationships should make this quite clear; the energy available, Q_{β^-}, is then greater by I^+/c^2, being equal to $(M_A^Z - M_A^{Z+1})$ instead of $(M_A^Z - M_A^{Z+1} - I^+/c^2)$. In β^+ decay the emitted positron and some negative electron will *subsequently* annihilate with the emission of two annihilation quanta each of energy $h\nu = 0.511\,\text{MeV}\ (= m_e c^2)$.

There is, however, a further most important result that emerges from the condition for β^+ decay given in Eq. 5-25, namely that even if

$$M_A^{Z-1} < M_A^Z < M_A^{Z-1} + 2m_e,$$

a transition can still occur (i.e., if the mass of the parent atom is greater than the mass of its isobar* one *less* in atomic number but is *not greater*

---

* Isobars are nuclides with the same atomic-mass number and different atomic numbers. The process of β decay is an isobaric transition.

than the mass of that isobar *plus* the mass of two electrons). This transition, which occurs by the process of electron capture, shows how important it is that we consider nuclear transitions in terms of the *atom as a whole*, as exemplified by relations 5-23, 5-24 and 5-25. We have already referred to internal conversion as being the result of a nuclear transition that involves the whole atom. In that process, the energy from the transition of the daughter nucleus to its ground or intermediate energy state is handed to an extranuclear electron which appears as a so-called internal-conversion electron. The energy may also appear in the form of γ rays. However, the process of internal conversion is not considered to involve an intermediate γ ray, but to be a direct interaction between the nucleus and the extranuclear electron.

The process of electron capture is likewise an interaction between the nucleus and its extranuclear electrons. Because it happens more frequently with the electrons in the inner shells, it is sometimes referred to as K-electron capture or K capture. However, lesser numbers of L- and M-shell electrons are also captured. The term "electron capture," or sometimes "orbital electron capture," is therefore more appropriate. Electron capture is equivalent to the opposite of a neutron decaying into a proton. (That is, $p + e^- \rightarrow n + v$, which is the reverse of n $\rightarrow p + \beta^- + \overline{v}$. When we change sides $-v$ is equivalent to $+\overline{v}$, or the absorption of a neutrino is the same as the emission of an antineutrino.) One example of an electron-capturing nuclide is iron-55, which decays

$$^{55}_{26}\text{Fe} + e^- \rightarrow {}^{55}_{25}\text{Mn} + v. \tag{5-26}$$

As the manganese-55 is left without a K- or L-shell electron, it promptly emits an Auger electron, or x rays characteristic of manganese. This possibility of electron capture was first predicted by H. Yukawa and S. Sakata in 1936 and was observed experimentally by L. W. Alvarez in 1938.

Considering the energy balance in reaction 5-26, the masses, on the carbon-12 scale, of iron-55 and manganese-55 are respectively 54.938302 amu and 54.938054 amu. The balance for the reaction is thus 0.000248 amu, which is very much less than $2m_e$ ($= 0.001098$ amu). The iron-55 can therefore only decay by electron capture, and then the 5.9-keV K_α x ray of manganese is emitted. Sometimes, however, a β^+ emitter also decays by electron capture, and there is a definite ratio between the two processes. Thus zinc-65 decays 98.54 per cent by electron capture and 1.46 per cent by β^+ emission. Many electron-capture events also leave the product nucleus in an excited state so that it emits a γ ray. A very low intensity of inner bremsstrahlung has also been observed in electron-capturing nuclides which have no accompanying γ rays. Such inner bremsstrahlung are present in all the different processes of β decay, since they arise from the acceleration or deceleration of electrons. The intensity of the inner bremsstrahlung is, however, so low that they are very difficult to observe in the presence of γ radiation. Their spectrum of energies is continuous up to the total energy of the nuclear transition involved.

Returning briefly to consider the energetics of the process of electron capture, we have

$$Z^A \rightarrow (Z-1)^{A*} + \nu,$$

where the daughter atom is left in an excited state, having lost an inner-shell electron and being left with excited outer-shell electrons. It readjusts, however, by emitting a series of x rays or other photons, so that it loses energy which we can designate by $\Sigma(h\nu)$. Therefore

$$M_A^{Z-1} = (M_A^{Z-1})^* - \Sigma(h\nu)/c^2,$$

and the energy available for the transition is

$$Q_{ec} = M_A^Z - (M_A^{Z-1})^*$$
$$= M_A^Z - M_A^{Z-1} - \Sigma(h\nu)/c^2.$$

Electron capture is therefore possible if

$$M_A^Z > M_A^{Z-1} + \Sigma(h\nu)/c^2.$$

The term $\Sigma(h\nu)/c^2$ is usually small compared with the difference in the masses. In general, therefore, electron capture can occur if

$$M_A^Z > M_A^{Z-1}. \tag{5-27}$$

It is interesting to note that for K-shell capture, or for each of the other shells (with or without γ-ray emission), the neutrinos must be monoenergetic $[\sim (M_A^Z - M_A^{Z-1})c^2$, when no γ ray is emitted].

Before leaving the subject of β decay it would be instructive to look at the decay of copper-64, which proceeds by electron capture and, as we have noted before, by both β^+ and β^- emission. The neighbouring isobars are nickel-64 and zinc-64, and the reactions involved are therefore

$$^{64}_{29}\text{Cu} \rightarrow {}^{64}_{30}\text{Zn} + \beta^- + \bar{\nu}, \qquad (5\text{-}28)$$

$$^{64}_{29}\text{Cu} \rightarrow {}^{64}_{28}\text{Ni} + \beta^+ + \nu, \qquad (5\text{-}29)$$

$$^{64}_{29}\text{Cu} + e^- \rightarrow {}^{64}_{28}\text{Ni} + \nu. \qquad (5\text{-}30)$$

The mass of copper-64 on the carbon-12 scale is 63.929761 amu, while the masses of zinc-64 and nickel-64 are respectively 63.929146 and 63.927959 amu. Thus copper-64 is heavier than both its isobaric neighbours in Z and is also heavier than nickel-64 by 0.001802 amu which is greater than $2m_e$ ($= 0.001098$ amu). Thus all three modes of decay are permitted.

References

Alvarez, L. W. (1938) The capture of orbital electrons by nuclei, *Phys. Rev.*, 54, 486.

Bethe, H. A. (1935) Masses of light atoms from transmutation data, *Phys. Rev.*, 47, 633.

Bethe, H. A. (1939) Energy production in stars, *Phys. Rev.*, 55, 434.

Chadwick, J. and Goldhaber, M. (1935) The nuclear photoelectric effect, *Proc. Roy. Soc.*, A, 151, 479.

Cockcroft, J. D. and Walton, E. T. S. (1932) Experiments with high velocity positive ions. II. The disintegration of elements by high velocity protons, *Proc. Roy. Soc.*, A, 137, 229.

Curie, Irène and Joliot, F. (1934) Un nouveau type de radioactivité, *Comptes Rendus, Académie des Sciences*, 198, 254.

Fermi, E., Amaldi, E., D'Agostino, D., Pontecorvo, B., Rasetti, F. and Segré, E. (1935) Artificial radioactivity produced by neutron bombardment-II. *Proc. Roy. Soc.*, A, 149, 522.

Livingood, J. J. (1936) Deuteron-induced radioactivities, *Phys. Rev.*, 50, 425.

Moon, P. B. and Tillman, J. R. (1936) Neutrons of thermal energies, *Proc. Roy. Soc.*, A, 153, 476.

Oliphant, M. L. E., Kempton, A. R. and Rutherford, E. (1935) Some nuclear transformations of beryllium and boron, and the masses of the light elements, *Proc. Roy. Soc.*, A, 150, 241.

Rutherford, E. (1919) Collision of α particles with light atoms. IV. An anomalous effect in nitrogen, *Phil. Mag.*, 37 [6], 581.

6 *Radiation Detectors*

THE previous chapters have been concerned with the energetics of nuclear change, the interactions of radiation with matter and developments in the field of radioactivity, treated largely in historical context, since its discovery in 1896. The closing chapters are devoted to present-day methods used both to measure activity (a quantity measured in units of reciprocal time) and to determine the nature of the radiations emitted in radioactive transitions (i.e. whether they be charged particles or electromagnetic radiations, the energies associated with such radiations, and their probabilities of emission per radioactive transition, or decay).

All methods of detection of charged-particle or electromagnetic radiations depend on the interactions of such radiations with the matter that they traverse and to which they impart energy by the ionization or excitation of its constituent atoms or molecules. Radioactivity measuring systems therefore consist essentially of (i) a radiation detector, or transducer, in which radiation energy is converted into electrical, optical or thermal energy or into energy of chemical change, and (ii) electrical, electronic, or other instrumentation by means of which the magnitudes of these effects in detectors may be determined. The earliest, and still often used, detector of radioactivity was the photographic emulsion by means of which chemical reactions caused by radioactivity are recorded, and such was the means of detection that led Becquerel to the discovery of radioactivity.

The rest of this chapter will be concerned with describing the physics of radiation detectors currently used for measurements of radioactivity. This will be followed by a chapter describing the associated electronic instrumentation (often described as "nucleonics"), and a concluding chapter devoted to a description of a number of radioactivity measuring procedures selected chiefly for their applicability to the assay of radiopharmaceuticals in nuclear-medicine laboratories.

Radiation detectors are based on the interaction of radiation with matter to produce ionization in gases, scintillations in solids, liquids and gases, liberated valence electrons and corresponding vacancies ("holes") in semi-conducting crystals, and temperature increases in the thermal elements of calorimeters ("thermels") due to the radiation energy deposited therein. The underlying physical principles of each will now be described.

Gas Ionization Detectors

Consider a volume of a pure gas within an enclosure containing two insulated electrodes, an anode and cathode, by means of which an electric field can be applied across most, if not all, of the gas volume. If any kind of ionizing radiation traverses the gas contained between the electrodes, it will dissipate its energy by interactions with the gas atoms or molecules, producing atomic or molecular excitation and ionization (see Chapter 3). This results in the formation of positive-negative ion pairs, usually consisting of atomic or molecular positive ions and electrons. In some gases, the so-called electronegative gases of which oxygen is an example, the electrons may become attached to neutral molecules to form negative ions. In general, the use of such electronegative gases is to be avoided and, unless specifically stated otherwise, the term "ion pair" will signify a positive atomic or molecular ion and a negative free electron. If oxygen is present, as in the case of air, negative molecular ions will, however, also be formed. When radiation traverses a volume of air the average energy expended in the formation of one ion pair, $\bar{W}_{air}$, is equal to approximately 34 eV. This energy expenditure also includes that expended in processes other than ionization. The average energy, $\bar{W}$, expended by alpha or beta particles or gamma rays in any other typical gas used as a filling in a gas ionization detector is of the order of 25 to 35 eV. Thus a 5-MeV alpha particle will form of the order of 1.5×10^5 ion pairs in expending all its energy in a gas.

The positive ions and electrons will now move under the influence of the applied electric field towards the cathode and anode, respectively. The ionization current so developed is utilized in three types of radiation detector, namely ionization chambers (which can be operated in the direct-current and current- or voltage-pulse modes), proportional counters and Geiger-Müller counters.

Direct-Current Ionization Chamber

An ionization chamber is an instrument, typically cylindrical in shape, with two electrodes so that an electric field may be applied across the volume of gas within the chamber. Such a chamber is illustrated in Fig. 6-1, together with a simple current-measuring circuit. Charged particles

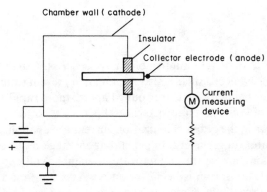

FIG. 6-1   Cylindrical gas ionization chamber with external circuit.

and electromagnetic radiation traversing this gas undergo inelastic collisions with atoms or molecules, ionizing them and thereby forming positive ions and electrons, which, in the absence of an electric field, will recombine to form neutral atoms.

When a voltage is applied to the electrodes, the electrons and ions drift along the lines of force, thus producing an *ionization current*. At STP, electrons drift at speeds of the order of 10^6 cm s^{-1}, while the drift of ions can be as many as three orders of magnitude slower, as their more frequent collisions with gas molecules prevent their attaining higher velocities. When an ionization chamber is subjected to a constant source of ionizing radiation, the measured current first increases with increasing voltage and then levels off. In the region of lower voltage recombination of

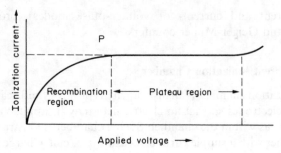

FIG. 6-2   Saturation curve (ionization current *vs.* applied voltage) for a typical ionization chamber.

the ions occurs, that is, the number of ions *collected* is less than the number *formed*. At P, the electric field is strong enough to prevent recombination; the ions, after being formed, are "pulled apart" more rapidly, and their probability of collection approaches 100 per cent.

The *plateau* in the graph of ionization current *versus* applied voltage is called the saturation current. In general, the voltage required to attain this saturation current for any ionization chamber will depend on the rate at which ionization is being produced. At saturation, the average ionization current i is related to the average number $\overline{N}$ of ion pairs produced per unit time, by the equation

$$i = \overline{N}e,   (6-1)$$

where e is the elementary charge. Each ion pair transfers $1.602 \times 10^{-19}\,C$ and the collection of 6.25×10^6 ion pairs $(10^{-12}\,C)$ per second corresponds to a current of $10^{-12}\,A$, or $1\,pA$. Thus the ionization-chamber current is a measure of the integrated or total effect of a large number of ionizing events. On the average, approximately $34\,eV$ are required to produce one ion pair in air, and this holds true for α and β particles as well as for x and γ rays.

The time of response of the current-detecting device is generally made long to suppress statistical fluctuations, and whether the electrons are collected as free electrons (as in argon) or, attached to slow-moving molecules, as negative ions (as in oxygen) is of no importance. It should be noted that the probability of recombination is lower in a gas in which only free electrons are formed, and saturation can be reached with a lower field intensity.

When the ionization current is sufficiently large, it may be measured with a microammeter. In general, however, it is necessary to use more sensitive methods, as, for example, by allowing the charge to collect on the total capacitance of the chamber and an external capacitor, and measuring the resulting rate of change of voltage on this capacitance with an electrometer system. Thus, since $Q = CV$ (where Q is the charge in coulombs, C is the total capacitance in farads, and V is the potential difference in volts), it follows that the current, i, may be determined by the relation

$$i = \frac{dQ}{dt} = C\frac{dV}{dt} \qquad (6\text{-}2)$$

In order to measure the voltage across the capacitor it is necessary to use a very high-impedance electrometer system such as that illustrated in Fig. 1-1. Solid-state versions are now, however, available. Sensitive methods of current measurement will be described in Chapter 7.

The number of ion pairs formed, per unit length of path, by a primary ionizing photon or charged particle traversing an ionization chamber is a function of the density of the gas filling the chamber. Therefore, in order to obtain reproducible results in unsealed chambers containing air, the ionization current will require correction to standard temperature and pressure (STP) when the range of the primary radiation exceeds the dimension of the chamber in the direction travelled. This correction is not needed with the pressure ionization chambers now in frequent use, which usually contain argon up to pressures of 20 atmospheres or more.

For the assay of gamma-ray solution or gaseous sources contained in glass ampoules, the solid-rod anode of Fig. 6-1 will be replaced with a hollow, re-entrant, metal anode with a vertically disposed axis, into which the ampoules can be placed. Windowless ionization chambers for the measurement of solid alpha-particle and beta-particle sources deposited on discs or planchettes are also used. For further details the reader is referred to the bibliography at the end of this chapter.

Gold-Leaf and Lauritsen Electroscopes

The gold-leaf electroscope was probably one of the earliest forms of ionization chamber. It was first described by Abraham Bennet in the *Philosophical Transactions of the Royal Society* of 1787 and was based on the earlier pith-ball electroscope of T. Cavallo (1770) which, in turn, probably derived from the electrical repulsion of two linen threads used by Benjamin Franklin (1706–1790) and the electrical needle of William Gilbert (1544–1603). It is of interest to remember that the word "electricity," coined by William Gilbert in the XVIth century, was derived from the Greek "$\eta\lambda\varepsilon\kappa\tau\rho o\nu$" (electron) for amber, which fossilized resin had the property, when rubbed, of attracting very light particles. It was not, however, until the discoveries of Röntgen and Becquerel, in 1895 and 1896 respectively, that the parallel-plate ionization chamber (Fig. 1-1) and gold-leaf electroscope (Fig. 1-3) became the accepted detectors of ionizing radiation and continued in this rôle for at least a decade.

The gold-leaf electroscope consists essentially of two rectangular gold leaves hanging from an insulated metal rod in a grounded case. A variant design is one gold leaf attached to a metal rod that terminates in a rigid metal strip (Fig. 1-3). The metal rod is usually terminated outside the case by a metal plate or sphere. On charging the leaf-system to a potential above or below that of the case, the leaves repel each other and diverge. On exposing the electroscope to ionizing radiation the leaves collapse as they are discharged by ion pairs that separate and move towards the leaves and case under the influence of the electric field that exists between the leaf system and the case. Their time rate of collapse can be used as a measure of the intensity of the ionizing radiation. For more than sixty years the electroscope has been the instrument of choice for the calibration of radium sources in national laboratories. For such measurements the radium source is usually supported on a modified optical bench at reproducible distances from the electroscope.

The Lauritsen electroscope is the modern and elegant version of the gold-leaf electroscope and its design is illustrated in the sectioned drawing in Fig. 6-3. This instrument was designed and constructed in the 1930's by Charles and Thomas Lauritsen, and it operates on the same principle as the gold-leaf electroscope, with the thin gold leaves replaced

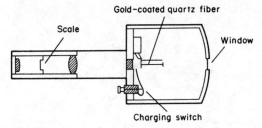

FIG. 6-3   Cross sectional view of a Lauritsen electroscope.

by a small metal frame to which is attached a fine gold-coated quartz fiber, 0.003 mm in diameter and 6 mm in length. To the free end of this fiber there is fastened, orthogonally, a very short length of the 0.003-mm-diameter fiber, so that its movement can be easily followed across the scale of a low-power microscope (Fig. 6-3). This assembly is mounted inside, and electrically insulated from, a cylindrical aluminum can, which serves as one of the electrodes of an ionization chamber. The frame and fiber represent the other electrode. To commence electroscope operation, the electrodes are momentarily connected to a positive voltage of, say, 100 V (the inventors charged the instrument with a "comb or fountain pen"), which causes the fibre to deflect away from the frame. In the presence of ionizing radiation, the negative ions and free electrons in the chamber will be attracted to the fiber system and will neutralize the charge thereon, allowing the elastic force in the quartz to return the fiber to its uncharged position.

Because the restoring force of the quartz fiber is due to its own elasticity and because the fiber is so light in weight that it is not affected by gravity, the Lauritsen electroscope is much more versatile than the gold-leaf instrument, as it can be mounted in any position. Its simplicity makes it a very reliable instrument, but its usefulness is frequently overlooked when routine measurements of radioactivity are performed. Gamma-ray sources are usually supported in a jig at reproducible distances from the electroscope for measurement. A modification of the Lauritsen electroscope, for the insertion of solid sources of alpha-particle and beta-particle-emitting radionuclides into the electroscope, is shown in Fig. 6-4. With the Lauritsen electroscope, relative measurements of activity can be made with reproducibilities of 1 or 2 per cent. This

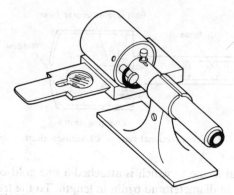

FIG. 6-4 Modified Lauritsen electroscope for the internal measurement of charged-particle-emitting radionuclides (from R. F. Hunter and W. B. Mann, National Research Council of Canada Report CRM-409, 1948).

instrument was one of the most widely used in the 1930's at the Lawrence Berkeley Laboratory for the identification and half-life measurements of new radionuclides produced on the 30-inch cyclotron. It was also used to assay samples of ^{24}Na and ^{32}P used in many of the early experiments in nuclear biology and nuclear medicine.

Quartz-fiber electrometers were also incorporated into the internal-source alpha-particle and beta-particle, and external-source gamma-ray ionization chambers developed at the Chalk River Laboratory of the National Research Council of Canada by Hugh Carmichael in the late 1940's.

Quartz-fiber electrometers are also the basis of the widely used direct-reading pocket dosimeters.

Pulse Ionization Chambers

The pulse mode of ionization-chamber operation is that of choice when the time rate of occurrence of the ionizing events is too low for convenient direct-current measurement, or when it is desired to determine either the rate of occurrence of ionizing events in the chamber, or the energy distribution of charged heavy particles that transfer all their energy to the filling gas in the sensitive volume of the chamber. In the last application the ionization chamber is functioning as a particle-energy spectrometer. Ionization chambers operated in the pulse mode are normally used for the detection of heavy ionizing particles such as alpha particles or fission fragments, and may be of either parallel-plate geometry (e.g. the Frisch-grid chamber), or of cylindrical geometry (Fig. 6-1) such as have found frequent use in the monitoring of radon concentrations in the air in uranium mines, by measuring the alpha-particle activities of air samples within the chamber.

Consider the parallel-plate ionization chamber, illustrated in Fig. 6-5, in which the plates usually have a large area in comparison with their

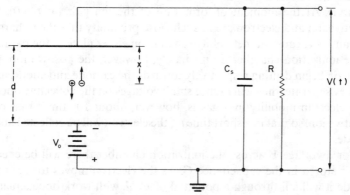

FIG. 6-5   Parallel-plate ionization chamber in which a single ion pair is assumed to have been formed at $t = 0$.

separation d. The high-voltage plate (cathode) of the chamber is connected to the negative pole of a source of electric potential and the collecting plate (anode) is connected to the positive pole of the source of potential, and to ground, through a high resistance R shunted by a capacitor of capacitance C_s. The output voltage signal $V(t)$, which varies functionally with the time t, is measured between the anode collector and ground, across the resistance R. In the absence of ionization in the chamber the collecting anode will remain at ground potential and $V(t) = 0$.

If, however, an alpha particle traverses the chamber, it will create an average number, $\bar{N}$, of ion pairs as a result of which charges of $+\bar{N}e$ and $-\bar{N}e$ will be swept towards the cathode and anode respectively. For an alpha particle of energy 5 MeV, $\bar{N}$ will be of the order of 1.5×10^5 ion pairs, and R is, in practice, made so large that, in the short time required to collect the electrons at the anode, the current flow in the external circuit can be neglected. It is of interest to understand, at least qualitatively, how, under these conditions, the voltage pulse, $V(t)$, develops as a function of time, t.

Now let us again consider Fig. 6-5 and suppose that at time $t = 0$ an ion pair (one positive ion and one electron) has been formed at a distance x from the collecting anode which is separated by a distance d from the cathode. At the moment of formation of the ion pair (at $t = 0$), the positive ion and electron are in such close proximity that their charges, $+e$ and $-e$, cause no net voltage change on either plate of the parallel-plate ionization chamber. Immediately, however, the positive ion and electron begin drifting respectively towards the cathode and anode, and, as they separate, they will induce small voltages on the collecting anode. The electron mobility in a gas is, however, about 10^3 times that of a positive ion, so that in a short time t_1, the electron will be collected at the anode.

For a voltage V across the ionization chamber there will be energy $\frac{1}{2}CV^2$ stored in the capacitance C. As the electron moves towards the anode it will fall through a potential of xV/d, with work done equal to xeV/d. If, as a result, there is a voltage pulse of $-\Delta V_-$ on the collecting anode there will be a decrease in the energy stored in the capacitance C equal to $\frac{1}{2}C(V - \Delta V_-)^2 - \frac{1}{2}CV^2$, which is approximately equal to $-CV \cdot \Delta V_-$ (neglecting the term ΔV_-^2). But the decrease in energy must equal the work done on the electron, or

$$-CV \cdot \Delta V_- = xeV/d,$$

or

$$\Delta V_- = -xe/dC. \tag{6-3}$$

During the time t_1, the positive ion will, however, have only moved about one thousandth of the distance travelled by the electron, so that the total voltage change on the anode will be essentially $-\Delta V_-$. In the additional time $t_2 - t_1$ the positive ion will move the distance $d - x$ to be neutralized at the cathode, with work done equal to $(d - x)eV/d$, and a corresponding voltage pulse, $-\Delta V_+$, on the anode given by

$$\Delta V_+ = -(d - x)eV/dC. \qquad (6\text{-}4)$$

The total voltage pulse on the anode is ΔV, where

$$\Delta V = \Delta V_- + \Delta V_+,$$

or $$\Delta V = -e/C. \qquad (6\text{-}5)$$

If, in Eqs. 6-3 and 6-4, $x = d - x$, or $x = d/2$, then $\Delta V_- = \Delta V_+$. That is, if the ion pair is formed half-way between the plates, the contribution to the output pulse by the electron is equal to that by the positive ion.

If, however, instead of a single ion pair we have $\overline{N}$ ion pairs formed in the ionization track of an alpha particle, the final voltage change on the collector at time t_2 will be $-\overline{N}e/C$. In the case of a single ion pair there will be a fairly sharp rise in negative potential of the collector plate between $t = 0$ and $t = t_1$, followed by a much slower increase in negative potential between $t = t_1$ and $t = t_2$. Ideally for a single ion pair these changes can be represented approximately linearly as shown in Fig. 6-6, with a discontinuity at t_1. But for $\overline{N}$ pairs there are two major moderating parameters, first the fact that an alpha-particle track will not

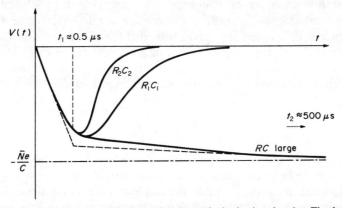

FIG. 6-6 Typical pulse outputs for a parallel-plate pulse ionization chamber. The dashed line is for a single ion pair and the solid lines, for $\overline{N}$ ion pairs. $RC \gg R_1C_1 > R_2C_2$.

necessarily be parallel to the collecting plate and, secondly, that $\overline{N}$ is only an average value and that the actual numbers of ion pairs formed in individual alpha-particle tracks will fluctuate statistically around the average number $\overline{N}$. The resultant voltage-pulse shape will therefore be smoothed out in the manner shown by the solid curve, for large RC, in Fig. 6-6 and approaches asymptotically to $-\overline{N}e/C$ as t approaches t_2.

In the $V(t)$-versus-time curve of Fig. 6-6, $V(t)$ is shown approaching asymptotically to the value $-\overline{N}e/C$, but this would only happen if the value of R were infinite. In actual fact, if a voltage $V(0)$ were to be suddenly imparted at time $t = 0$ to the collector system shown in Fig. 6.5, the value of $V(t)$ at any subsequent time t can be simply shown to be given by

$$V(t) = V(0)e^{-t/RC}$$

The product of RC is known as the *time constant* of the circuit and is clearly equal to the time taken for the voltage $V(t) = -\overline{N}e/C$, to decay to $1/e$ of its initial value. By suitable choice of the time constant, $V(t)$ can be made to return rapidly to zero, as indicated in Fig. 6-6, and the resultant output pulse from the pulse ionization chamber can then be handled by additional electronic instrumentation for further shaping, amplification or pulse-height discrimination.

The process of formation of the pulse on the collector system of an ionization chamber of cylindrical geometry (Fig. 6-1) is similar in principle to that in a parallel-plate geometry. In the latter, however, the electric field between the plates, provided the plates are large in area compared with their separation, is uniform whereas in the cylindrical chamber the field gradient increases rather rapidly near the rod anode, the electric intensity E_r at radius r being given by the relation

$$E_r = \frac{V}{r \ln b/a} \tag{6-6}$$

where a is the radius of the anode, b is the radius of the cathode, and V is the voltage between them. For an alpha particle moving through the cylindrical chamber parallel to the anode and close to it, the collection time for the electrons relative to that for the positive ions will therefore be much shorter than in the case of the parallel-plate chamber, for the same relative distance of the alpha-particle track between the anode and cathode.

The Frisch-Grid Ionization Chamber

Let us consider the output pulses of a parallel-plate ionization chamber in which an alpha-particle source is located centrally on the lower cathode plate (Fig. 6-7). The value of the time constant, RC, is also

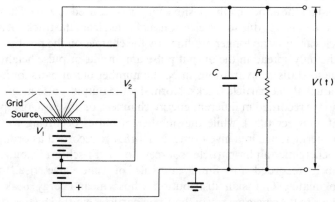

FIG. 6-7   Schematic of Frisch-grid ionization chamber with α-particle-emitting source.

assumed to be such as to give rise to "clipped" output pulses such as those illustrated in Fig. 6-6 for the shorter time constants R_1C_1 or R_2C_2. For any chamber of typical size the electron contribution to the pulse will be collected in the order of 1 μs, wherever the electrons in the alpha-particle track are located. In this time, however, the positive ions will have moved only of the order of one-thousandth the distance moved by the electrons. The amplitude of the clipped pulse at approximately t, will therefore be very dependent upon the relative proximity of the positive ions to the anode because the pulse amplitude is equal to $-\overline{N}(e/C - \Delta V_+)$, and ΔV_+ (or strictly the average of ΔV_+ for all positive ions in the track) is a function of the orientation of the alpha-particle track. In order to use the parallel-plate ionization chamber to distinguish between different monoenergetic groups of alpha particles, i.e. as a spectrometer, it would, of course, be necessary to eliminate this dependence of the pulse amplitude on the direction of emission of the alpha particles from the source so as to obtain the best possible resolution. Otto Frisch achieved this by simply placing a fine wire grid parallel to the plates and beyond the range of the alpha particles, as shown in Fig. 6-7. The grid, which was highly transparent to the accelerated electrons, was maintained at a potential intermediate between the anode potential and ground, so that it shielded the anode collector from the cloud of positive ions. In this way the voltage, $\overline{N}\Delta V_+$, induced on the anode by the positive ions was reduced to zero. The pulse amplitudes, $\overline{N}e/C$, due to the electrons collected from the tracks of alpha particles are now no longer modified by the charges on the positive ions, and the only spread in the output pulse amplitude, or pulse height, will be due to statistical variations in N, the number of ion pairs formed in individual alpha-particle tracks. Normally the output pulse heights are nowadays recorded in different energy channels, or abscissa, of a pulse-height analyzer (q.v.), while the number of pulses occurring within a given energy range in a given time, $N(E)\,dE$, are recorded as ordinates. Monoenergetic alpha-particle sources in a Frisch-grid ionization chamber, recorded for finite intervals of time, give rise to an approximately Gaussian distribution whose mean-energy peak corresponds to the single energy of the alpha particles emitted by the source. The resolution of the alpha-particle peak, usually measured by the width of the peak at half the ordinate at the middle of the peak, is normally such that 5-MeV alpha particles can, for example, be resolved from monoenergetic groups of alpha particles having energies of approximately 4.95 MeV and 5.05 MeV.

For use as an alpha-particle-energy spectrometer the output pulse height of such a gridded ionization chamber is calibrated in terms of energy using a source that emits monoenergetic alpha particles of known energy.

Gas Multiplication

When the voltage applied across an ionization chamber containing a pure electropositive gas is continuously increased through the plateau region, corresponding to the saturation current (Fig. 6-2), the electrons from the ion pairs formed by the primary ionizing event begin to acquire sufficient energy between collisions with neutral gas atoms or molecules to cause secondary ionization. The ionization current, or the pulse height in the pulse mode of operation, will therefore start again to rise because of the formation of these additional ion pairs and the process of *gas multiplication* or *gas amplification* occurs. The plateau represents a region of no gas amplification. The process of gas multiplication will, however, occur at much lower values of the applied voltage in the case of the cylindrical geometry of Fig. 6-1 than in that of the parallel-plate ionization chamber (Fig. 6-5), because of the greatly enhanced electric-field intensity near the rod anode, as indicated by Eq. 6-6. If the rod anode in the cylindrical chamber is now replaced by a fine wire, the electrons will pass through fields of even greater electric intensity before collection on the wire because the value of the radius r in the gas, in close proximity to the wire, cannot be less than the radius, a, of the wire (Eq. 6-6). Thus for a given applied voltage V, the electric-field intensity, E_r at radius r, can be many times greater 0.02 mm from the surface of a fine wire than 0.02 mm from a rod. For example, for a rod and a wire with radii of the order of 0.2 cm and 0.002 cm, respectively, the ratio of electric-field intensities will be about 20, for a cylindrical cathode with radius b equal to 5 cm in both cases. With such a detector almost all the secondary ionization, and consequent gas multiplication, occurs in a very small volume of gas in close proximity to the anode wire. This multiplication is in the form of an avalanche, the electron of one ion pair forming another ion pair, and so on, in rapid succession in the intense electric field near the wire. Such an avalanche is often called a *Townsend avalanche*, after J. S. Townsend who first demonstrated the phenomenon in electrical discharges in gases. In a gas ionization detector with a fine-wire anode of the order of 0.002 cm the avalanche has been found to occur within about 0.005 cm from the surface of the anode wire.

Proportional and Geiger Regions

In order to understand the operation of gas ionization detectors in the proportional and Geiger regions let us consider, in such a detector of cylindrical geometry, the development of the pulse amplitude with increasing voltage, from the region of no gas multiplication at lower voltages, through gas multiplication, to the condition of continuous discharge which must occur at sufficiently high values of the applied voltage. This is illustrated in Fig. 6-8. The voltage V across the detector of cylindrical geometry will be applied by means of an external circuit similar to that shown in Fig. 6-5. The total capacitance C of the system is that due to the capacitance, C_s, across the resistance R, and the capacitance of the gas ionization detector itself.

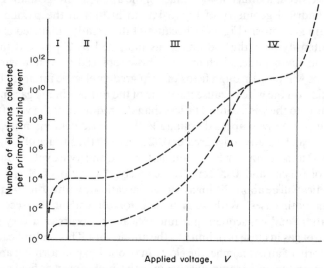

FIG. 6-8   Electrons collected *vs.* applied-voltage curves illustrating the recombination (I), saturation (II), proportional (III) and Geiger (IV) regions, for primary ionizing events forming respectively $N_1 = 10$ and $N_2 = 10^4$ ion pairs. Point A indicates the onset of limited proportionality (see text).

One might equally well consider the increase in ionization current with increasing voltage, but ionization detectors that employ gas multiplication are usually operated in the pulse mode in order to count individual events.

It is also still assumed that the gas filling the chamber is a pure electropositive gas so that the ion pairs formed by an ionizing event in the chamber are composed of positive ions and free electrons, rather than of positive and negative gaseous ions, that the gas pressure is approximately atmospheric, and that the anode consists of a fine wire so that any secondary ionization will occur in a region in very close proximity to the anode wire. We will consider, separately, two ionizing events one of which produces N_1 ion pairs and the other N_2, and by way of example, it will be assumed that N_1 is equal to 10 and N_2 to 10^4 ion pairs, which, for an average of 34 eV for the formation of one ion pair, would correspond to energy depositions—in the detector—of, respectively, 340 eV and 340 keV.

By some hypothetical mechanism we will assume that we are provided with a series of ionizing events that create either 10 or 10^4 ion pairs in the detector and that these are discrete events in that all the electrons and positive ions from each event will have been collected before the next one occurs. We will then look at the effect on the series of discrete output pulses, across the resistance R, of increasing the applied voltage V, for both $N_1 = 10$ and $N_2 = 10^4$ ion pairs.

Consider first the case of N_1 equal to 10 ion pairs. For $V = 0$ there will be no separation of the ion pairs, except by thermal agitation, and they will all almost certainly recombine, with none reaching either electrode of the detector. As the voltage is increased the probability of recombination decreases and an increasing number of electrons will reach the anode, and the voltage pulse across R ($= Q/C$ where Q is the charge collected) will increase with applied voltage in the manner shown in region I of Fig. 6-8. At the high-voltage limit of this region the applied voltage is sufficient to prevent any recombination and all 10 ion pairs will be collected. The resultant output voltage across R will approach asymptotically towards $10\,e/C$, for sufficiently large values of RC, but, depending on the value of RC, will, at some point, start returning to zero. In practice, however, the values of R and C will be such as to give the clipped pulses illustrated in Fig. 6-6, which fall short of $10\,e/C$ even though the 10 electrons are collected on the anode.

As the applied voltage is increased further the number of ion pairs collected remains equal to the number formed and we pass into region II (Fig. 6-8) which is known as the *saturation* region or ionization-chamber region. This region may extend up to applied voltages of several hundred volts depending on the physical dimensions of the detector and the nature and density of the filling gas. In the case of the more heavily ionizing event ($N_2 = 10^4$ ion pairs) the onset of the plateau in region II will occur at a slightly higher applied voltage, because at any given voltage the probability of recombination will be somewhat greater for larger numbers of ion pairs.

On increasing the voltage beyond the upper limit of region II a new phenomenon is observed, namely that of *gas multiplication*, in which the electrons receive sufficient energy between collisions to cause secondary ionization of the gas atoms or molecules with which they interact. The electrons from the ion pairs so formed can, as they are, in turn, accelerated towards the collecting anode, give rise to further successive generations of ion pairs and a Townsend avalanche is developed. This process of gas multiplication is also sometimes called *gas amplification* while, by contrast, region II is sometimes called the region of *zero gas amplification*. For lower values of the applied voltage in region III, where the ratio of the ordinates of the N_1 and N_2 curves remains constant for all values of the applied voltage, the charge collected on the anode is found, apart from statistical fluctuations, to be proportional to the energy deposited by the incident radiation, i.e. to the number of ion pairs formed in primary ionizing events. Region III is therefore known as the *proportional* region. The same proportionality holds also, of course, in the saturation region, II, of ionization-chamber operation. As is illustrated in Fig. 6-8, however, proportionality begins to break down for values of the applied voltage greater than that represented by the "dashed-line" ordinate, above which the ratio of the number of electrons collected at any given voltage, for curves N_1 and N_2 respectively, begins to fall below the ratio of $N_2 : N_1$. This subregion between the "dashed-line" ordinate and the upper limit of region III is known as the region of *limited proportionality*, which is somewhat of a euphemism because the N_1 and N_2 curves are rapidly converging towards each other and at A, the ratio of the ordinates, N_2 to N_1, is about 10 instead of 10^3. The number of electrons collected at the anode, and the resulting amplitude of the output voltage pulse, are no longer linear functions of the energy deposited in the detector by the primary ionizing radiation.

As the applied voltage is still further increased beyond the upper limit of region III, the N_1 and N_2 curves coalesce, the number of electrons collected at the anode becomes independent of the number of ion pairs formed by the primary ionizing event, and we enter region IV which is known as the *Geiger* region. In this region a primary ionizing event forming one or two ion pairs will generate an output voltage pulse with as large an amplitude as that generated by a primary ionizing event that forms 10^4 ion pairs.

In the proportional region the Townsend avalanches near the anode wire are rather limited in size and do not extend too much further along the wire than the projection upon the wire of the primary ionizing track. In the region of limited proportionality, gas multiplication by factors in excess of 10^3 takes place and two other physical effects begin to predominate, namely the space charge effects due to large numbers of electrons and positive ions, and a spreading of the Townsend avalanches along the anode wire caused by what might be called *photon multiplication.*

In the region of limited proportionality, and even more so in the Geiger region, the electric field near the anode becomes modified by the space charge due to the positive ions. As electrons reach the multiplicative region of high electric intensity near the anode wire first, second and subsequent generations of ion pairs are formed. For every electron formed in the primary ionizing event there may now be two, four, or many more pairs of electrons speeding towards the anode wire, creating a Townsend avalanche and leaving the positive ions more or less in the same positions where they were formed, their mobilities being about one thousand times less than that of the electrons. The cloud of positive ions that has been left behind will form a positive space charge that will be larger the higher the voltage applied across the detector because of the greater gas multiplication, and the space charge so created will tend to weaken the electric field near the wire for those electrons arriving late from those parts of the primary ionizing track furthest away from the anode wire. The larger the number of ion pairs in the primary ionizing event, and the higher the gas multiplication, the greater will be the inhibiting effect of the positive-ion space charge. Consequently the N_2 curve will tend to increase less and less rapidly, or with more rapidly diminishing slope, than the N_1 curve, and true proportionality, represented by the constant ratio of the N_2 to N_1 ordinates, will be lost.

Another effect which becomes significant in the region of limited proportionality is that of atomic and molecular excitation and de-excitation, with accompanying photon emission. This increases rapidly with increasing voltage, and the photons, which are emitted isotropically, generate photoelectrons from the cathode-wall surface. These photoelectrons will generate new avalanches along the wire and a positive-ion sheath begins to spread along the wire. Photoelectrons that are emitted from the cathode, by photons from the avalanches near the wire or by the neutralization of the positive ions near its surface, can cause delayed spurious pulses, known as *after-pulses*. In the Geiger region the positive-ion sheath envelopes the whole anode wire.

The region of limited proportionality is thus seen as a transition region between that of true proportionality and the Geiger region, where there is no proportionality, and all output pulses are of essentially the same size and may amount to several volts. The region of limited proportionality is not normally used for pulse counting, but may prove useful for the detection of low-energy events at relatively high count rates.

The proportional region is probably that which is most frequently used for the operation of gas ionization detectors in the pulse-counting mode, because of the negligible positive-ion space-charge effects and the rapid recovery of the detector with its consequent potential for very high rates of counting.

The counting characteristics of detectors operating in the proportional and Geiger regions will be discussed in greater detail in the sections immediately following.

When the voltage applied across the gas ionization detector is raised beyond the upper limit of region IV, the detector starts first to develop multiple pulses, often unrelated to any primary ionizing events, and then to go into a state of continuous gaseous electrical discharge, such as that developed in the familiar tubular fluorescent lamp.

Proportional and Geiger-Müller Counters

In the previous sections, on gas multiplication and the proportional and Geiger regions, we have considered a gas ionization chamber filled only with a pure gas in which the negative charge is carried by electrons. Only in the case of the direct-current mode of operation of the ionization chamber in the saturation region has air been considered as a filling, because, in this non-recombining and non-multiplicative region, it is unimportant if most of the negative charge is carried by negative oxygen ions.

In order to improve the stability of pulse counters using the principle of gas multiplication, small amounts of polyatomic "quenching" gases are added to a pure electropositive diatomic or monatomic gas, such as argon. The polyatomic gas is chosen to have lower excitation and ionization potentials than the pure gas with which it is mixed, and its purpose is to prevent the emission of photoelectrons from the cathode. It is therefore very important to state specifically whether we are considering a filling of such a gas mixture or one of a pure electropositive gas alone, because there are important differences in the counter operation. One such important difference is that cited by D. R. Corson and P. R. Wilson, namely that photons of wavelengths typical of those created in Townsend avalanches near the anode wire are reduced by absorption to $1/e$ of their initial intensity in a distance of 1.2 mm in alcohol vapor at a pressure of 1 cm of mercury.

First we will consider what takes place when the counter with its fine-wire anode is filled with a pure gas such as argon.

A proportional counter is, as its name implies, a gas ionization detector that operates in the proportional region. Its anode usually consists of a straight or looped fine wire, having a diameter of the order of 0.002 cm, and the cathode, which normally constitutes the enclosure for the filling gas, can be of any shape to suit the experimental conditions. Most often, for the detection of alpha and beta particles or electrons, the counter is composed of two detectors each subtending an angle of 2π steradians to a source mounted on a thin organic plastic support. These are the so-called 4π counters and are generally constructed in one of three geometries, namely (i) cylindrical, with two fine, insulated, anode wires stretched parallel to the axis of the cylinder (Fig. 6-9), (ii) spherical, or hemispherical, with the anode wire in the form of a small loop (Fig. 6-10), or (iii) the so-called "pill-box" geometry consisting of a cylindrical cathode whose height is small compared with its diameter, with the anode wire stretched along a diameter of a circular section of the cylinder (Fig. 6-11). Such counters have, in the past, been used also as Geiger-Müller counters, but the faster operation in the proportional region makes that the region of choice for modern high-speed counting.

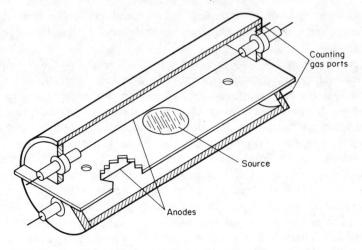

FIG. 6-9   Cut-away view of a 4π-cylindrical gas-flow proportional counter.

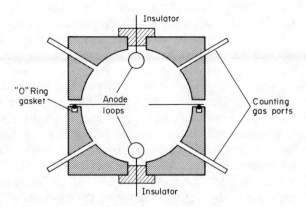

FIG. 6-10   Cross-sectional view of a spherical 4π gas-flow proportional counter.

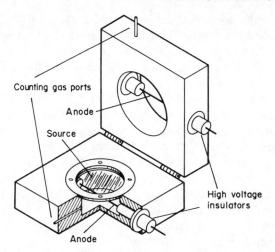

Counting gas ports

Anode

Source

High voltage insulators

Anode

FIG. 6-11   Cut-away view of a 4π "pill-box" gas-flow proportional counter.

The proportional counter is based on the principle of linear gas amplification that occurs in a region very close to the fine-wire anode where, as described in the last section, the number of ion pairs formed is directly proportional to the number of electrons created in the primary ionizing event. Because the avalanches that occur are formed in close proximity to the anode wire, the proportional counter, unlike the pulse ionization chamber, does not exhibit significant pulse-height dependence on the location, within the sensitive volume of the counter, of the primary ionizing event. The pulse height, or amplitude (Fig. 6-6), depends essentially on the numbers of electrons and positive ions moving respectively, towards and away from the anode and these, in turn, are related to the number of electrons, formed in the primary ionizing event, by the multiplication factor M.

In the acceleration of the electrons towards the anode and their transfer of energy to neutral gas atoms or molecules in inelastic collisions, both ion pairs and ultraviolet photons will be produced. These photons can, in turn, produce photoelectrons from the counter walls in the case of a pure gas or from molecules the gas additives if they have ionization potentials lower in energy than the energy of the photons. Using a somewhat simplified model, it is possible to derive an expression for M in terms of the average number, $\bar{m}$, of ion pairs formed, per electron entering the multiplication region near the wire, and the probability P, per secondary ion pair formed in the avalanche, of producing a photoelectron.

If n electrons are produced in the ion-pair track formed by the primary ionizing event, an average of $n\bar{m}$ secondary ion pairs and $n\bar{m}P$ photoelectrons will be produced in the Townsend avalanche resulting from the passage of the n primary electrons through the high-intensity electric field near the anode wire. The photoelectrons will now also be accelerated towards the anode and, in the multiplication region near the wire, will produce $n\bar{m}^2P$ more ion pairs and $n\bar{m}^2P^2$ further photoelectrons, and so on.

Thus each primary ionization track of n ion pairs will result in a total of N electrons reaching the anode wire, where

$$N = n\bar{m} + n\bar{m}^2P + n\bar{m}^3P^2 + \cdots$$

$$= n\bar{m}(1 + \bar{m}P + \bar{m}^2P^2 + \cdots). \tag{6-7}$$

In a proportional counter $\bar{m}P$ is much less than one, however, so that the series in $\bar{m}P$ converges and can be represented by the binomial expansion of $(1 - \bar{m}P)^{-1}$, so that

$$N = n\bar{m}/(1 - \bar{m}P),$$

$$\text{or } M = \bar{m}/(1 - \bar{m}P), \tag{6-8}$$

where $M = N/n$ is the multiplication factor. It is seen from Eq. 6-8 that the total number, N, of electrons reaching the anode at any given applied voltage is proportional to the total number, n, of ion pairs formed in the track of the primary ionizing radiation. The lesser multiplication factor $\bar{m}$ that applied only to electrons initiating Townsend avalanches is, however, an average value and varies statistically from electron to electron formed in any one primary ionizing event. The output pulse heights will therefore also vary statistically, from one primary ionizing event to another.

The condition that $\bar{m}P$ be much less than one is fundamental to the principle of proportionality. If $\bar{m}P$ approaches unity, the series of Eq. 6-7 diverges and the multiplication factor increases rapidly. One effect of increasing the applied voltage is to increase $\bar{m}$, the average number of secondary ion pairs formed, per electron entering the high-intensity electric field near the anode wire, with consequent increase in M. Clearly M is equal to $\bar{m}$ for $\bar{m}P \ll 1$, and if $\bar{m}$ is also equal to 1, then, from Eq. 6-8, $M = 1$ which is the condition for ionization-current saturation.

Photoelectrons can also be produced from the cathode when the positive ions are neutralized there. This, as has been mentioned, can give rise to the phenomenon of after-pulsing as can also photons generated in the Townsend avalanche region near the anode wire in a pure gas. This problem is described and discussed by P. J. Campion in the reference on spurious pulses given in this chapter's bibliography. That part of it arising from positive-ion neutralization at the cathode can, however, be decreased by the use of suitable filling-gas mixtures, such as a noble gas mixed with a small quantity of a polyatomic gas. One such mixture that is in common use consists of 90 per cent argon and 10 per cent methane. Because the first ionization potential of methane is lower than that of argon, an ionized argon atom, in one of its many collisions as it proceeds towards the cathode, can acquire an electron from a methane molecule that can then de-excite by dissociation when it is electrically neutralized near the cathode, instead of by photon emission. This mechanism greatly increases the stability of counters operating in the proportional region.

Other aspects of the phenomenon of after-pulsing are discussed in the reference in the bibliography. If, however, a proportional counter is operated at moderate values of the applied voltage, after-pulses are not likely to make a contribution greater than 1 in 10^4 of the pulses arising from true primary ionizing events.

In a proportional counter of typical dimensions, the electrons are collected in the order of microseconds, and the positive ions in the order of milliseconds. With reasonably short RC clipping times (Fig. 6-6), however, and in the absence of large positive-ion space-charge effects, which occur at higher values of the applied voltage, a proportional counter can recover and be ready to record another ionizing event in the order of a microsecond.

In the proportional region, every ionizing event gives rise to an output pulse, although those of lowest energy may be lost in the "noise" of the subsequent stages of electronic amplification. If there is a constant rate of occurrence of primary ionizing events, then, throughout the region of strict proportionality, there will be, apart from dead-time losses, an equal output-pulse rate from the proportional counter. Thus on increasing the applied voltage through region III (Fig. 6-8) up to the region of limited proportionality, we will obtain a count-rate *vs.* applied-voltage plateau similar to that obtained in the saturation region of operation of the ionization chamber in the direct-current mode. In the saturation region the plateau derives, however, from the constancy of the *number of electrons* collected at the anode, whereas in the proportional counter the plateau derives from the constancy of the *number of output pulses* irrespective of the amount of gas multiplication.

If now the voltage applied to the counter is increased through the region of limited proportionality, the multiplication, $\bar{m}$, in the avalanche region near the wire becomes large, and the numbers of photons capable of producing photoelectrons generated in the first avalanche, $\bar{m}P$, will no longer be very small compared with unity. As $\bar{m}P$ approaches equality to unity, the series of Eq. 6-7 begins to diverge, many ion pairs and photoelectrons are generated and an electrical discharge occurs, and the mode of operation is now that of the Geiger region. It is now, however, very necessary to use a polyatomic quenching gas otherwise the counter will continue to develop multiple pulses or go into continuous discharge. The quenching gas used in the past was often ethyl alcohol for which the first ionization potential is 11.3 eV compared with that for argon of 15.7 eV.

If the counter had been filled with a pure gas such as argon, then on entering the Geiger region, photons from the Townsend avalanches close to the wire would have been propagated isotropically throughout the counter, generating photoelectrons from the cathode walls which in turn would generate new Townsend avalanches. The argon positive ions would also produce photons on being neutralized near the cathode, and these photons would generate more photoelectrons and more avalanches with a consequent continuous discharge or multiple pulsing reflecting the transit time of the positive ions across the counter.

With the addition of ethyl alcohol to the argon, to give a mixture of 10 per cent alcohol to 90 per cent argon (for a total pressure, usually, of about 10 cm of Hg), the behaviour of the counter in the Geiger region is completely changed. As already mentioned the photons will be so effectively absorbed by the alcohol molecules that the probability of any reaching the cathode is extremely small. Moreover, many of the photons emitted by argon atoms in the avalanches near the anode wire will have energies sufficient to ionize alcohol molecules, thereby creating more electrons and more avalanches. Because of the larger absorption of the isotropically propagated ultraviolet photons by the alcohol molecules, the photoelectrons that they produce will be in very close proximity to the anode wire, and a positive-ion sheath spreads along the wire with a velocity of the order of 10^7 cm s^{-1}, and this positive-ion sheath, in turn, reduces the electric-field intensity near the wire so that no further ionization takes place and the counter pulse is "quenched". Furthermore, an argon positive ion moving towards the cathode can, as with methane in the proportional counter, exchange its state of ionization with an alcohol molecule, with its lower first ionization potential. When the alcohol ion is subsequently neutralized near the cathode it de-excites by dissociation, as does methane, without emission of a photon energetic enough to produce a photoelectron from the cathode, and this possible cause of after-pulsing is thereby eliminated.

Such a counter operating in the Geiger region is known as a Geiger-Müller counter, after H. Geiger and W. Müller, although one of the earliest uses of such a device appears to have been by Rutherford and Geiger in 1908.

Apart from statistical fluctuations, the output pulses from a Geiger-Müller counter for a given applied voltage, are always the same size, whether the primary ionizing event produces one ion pair or a thousand. The output pulse can also be as large as several volts, being composed of as many as 10^9 to 10^{10} ion pairs in the avalanches that are generated along the anode wire. Because of the large positive-ion space charges that are left around the wire after the initial discharge, such counters remain inoperative for times ranging from 10 to 500 microseconds, depending on the size and operating conditions of the counter. For a typical Geiger-Müller counter the dead time is of the order of 300 μs. They cannot therefore be used at nearly as high count rates as can proportional counters. The time required for a Geiger-Müller counter to becomes operable again is, however, considerably less than the time required to collect all the positive ions (see Chapter 7 for a discussion of counter dead times).

Self-quenched Geiger-Müller counters use small quantities of additives such as ethyl alcohol, ethyl formate, or a halogen such as bromine or chlorine. The organic-quenched counters have a limited life, of the order of 10^8 to 10^{10} counts, because the organic molecular ions are dissociated on being neutralized at the cathode (of the order of 10^9 to 10^{10} will be dissociated for each count). The organic-quenched counters have, however, better plateaux than the halogen-quenched counters. The halogen-quenched counters have essentially unlimited life as molecular ions dissociated at the cathode will quickly recombine, but they are more difficult to construct, as only very small and critical concentrations of the halogen are added (e.g., 0.1 per cent of chlorine in neon), and one must take great care to use materials that will not alter the concentration by absorption or chemical reaction with the highly active vapors.

In *non-self-quenched* Geiger-Müller counters, the counter is filled with a pure monatomic or diatomic electropositive gas. It is therefore necessary to employ a simple electronic external quenching circuit, whereby, immediately following a counter discharge, the voltage across the counter is reduced for a period of time much longer than the time required for the collection of the positive-ion sheath. During this interval of time the voltage across the counter is too low for photoelectrons ejected from the cathode to cause further gas multiplication, with consequent Townsend avalanches. Such external quenching is not much used at the present time.

Two of the gas ionization detectors that find widespread use today are the gas ionization chamber, increasingly in the form of the so-called "dose calibrator" used for the assay of the radiopharmaceuticals, and the proportional counter which, because of its short dead, or inoperable, time can be used to count, with special circuitry, radiation events at rates up to about 10^6 s^{-1}! The practical application of both these detectors will be discussed in Chapter 8. Before leaving this subject, however, it is of interest to consider the relative roles of electrons and positive ions in the formation of the output pulse of a proportional counter. The development of the pulse in a parallel-plate ionization chamber has been schematically illustrated in Fig. 6-6, and that for a cylindrical proportional counter, the operation of which depends on gas multiplication, is shown in Fig. 6-12.

The small increase in the negative potential between time zero and t_1, shown in Fig. 6-12, represents the time required for the electrons from the primary ionizing event to reach the multiplication region. The

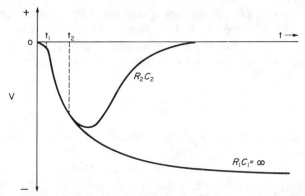

FIG. 6-12   Development of a pulse from cylindrical gas proportional counter. $R_1C_1 > R_2C_2$.

electrons from such an event occurring at the order of 1 cm from the anode wire, with drift velocities of the order of 10^6 cm s^{-1}, would take about 1 microsecond to reach the multiplication region which is within about 0.002 cm from the wire. From time t_1, the negative voltage pulse rises very fast, first as a result of the movement of the electrons towards the wire in the interval of time from t_1 to t_2, and then as a result of the movement of the positive ions outwards in the strong field near the wire. The rate of increase of the negative voltage pulse on the wire decreases as the positive ions reach the weaker regions of the electric field, and approaches asymptotically to zero as the positive ions approach the cathode and are then neutralized there. If the time constant RC is short, then the pulse will return rapidly to zero in the manner shown in Fig. 6-12. The time $t_2 - t_1$ can be of the order of 2×10^{-9} second and the time scale in Fig. 6-12 is grossly distorted. The initial pulse rise is, however, mainly due to the movement of the positive ions, as may be seen from the following qualitative description.

In the uniform electric field of the parallel-plate ionization chamber operating in the pulse mode, the pulse results in about equal parts from the movement of the electrons and of the positive ions, when the primary ionization is produced at a point equidistant between the electrodes (Fig. 6-5). In the cylindrical proportional counter the electric field intensity rises very rapidly in the region close to the anode wire, and intense ion-pair formation occurs within one or two thousandths of a centimeter from the surface of the wire. For a cylindrical counter of 2-cm radius having an anode wire of 0.002-cm radius, 10 per cent of the drop in voltage across the counter occurs within a distance of 0.002 cm from the surface of the anode wire. This is readily seen by integrating Eq. 6-6 to give

$$V_r = \frac{V}{\ln b/a} \ln r/a \qquad (6\text{-}9)$$

where V_r is the electric potential at radius r from the axis of the counter. Substituting $b/a = 1000$ and $r/a = 2$ gives $V_r = V/10$.

If, as before, C is the total capacitance across the counter and V is the applied voltage then the energy stored in the system is $\frac{1}{2}CV^2$. If now, one ion pair is formed at a distance of 0.002 cm from the wire (i.e. at $r = 0.004$ cm), the electron will move through a potential difference of $V/10$ and the positive ion through a potential difference of $9V/10$, and the work done by each will be respectively $eV/10$ and $9eV/10$. These amounts of work done will result in corresponding voltage pulses ΔV^- and ΔV^+, and increases in the energy stored in the system. The energy increase due to the electron will occur in about $0.002\,\mu s$ and will be equal to $\frac{1}{2}C(V + \Delta V^-)^2 - \frac{1}{2}CV^2$, which is approximately equal to $CV \cdot \Delta V^-$, whence

$$CV \cdot \Delta V^- = eV/10,$$

or
$$\Delta V^- = e/10C \qquad (6\text{-}10)$$

During approximately the next $2\,\mu s$, during which the positive ion travels the 2 cm to the cathode, its contribution to the pulse, ΔV^+, will develop and as it reaches the cathode

$$\Delta V^+ = 9e/10C \qquad (6\text{-}11)$$

From Eqs. 6-10 and 6-11 we see that the total pulse, $\Delta V^- + \Delta V^+$, due to the formation of the single ion pair will be equal to e/C, to which the electron will have only contributed 10 per cent. Actually this calculation was based on formation of a single ion pair at a radius, r, equal to 0.004 cm, but the center of gravity of each avalanche is likely to be even closer to the wire, and the electron contribution will be correspondingly less than 10 per cent. Thus $t_2 - t_1$ in Fig. 6-12 would be much smaller than can be shown (between 10^{-2} and 10^{-3} of t_1), and the contribution to the pulse by the movement of the electrons may be between 0.1 and 10 per cent of that of the positive ions.

A variant of the gas ionization detector that no longer finds use today, except possibly as a lightning conductor, is the Geiger point counter that was described in 1913. This was cylindrical in geometry, and was made from a brass tube which contained the gas. The axial electrode was, however, a stiff rod supported at one end only, and terminating in a sharp point. The tube was maintained at a high positive potential, near to the discharging point, and the gas in the very intense electric field near the point required only weak ionization to trigger an electrical discharge, the magnitude of which was largely independent of the initial ionization. When the point was observed by means of a microscope, a sharp flash of light could be seen when an alpha particle entered the detector through a thin mica window. The pulses were counted by means of a string electrometer.

The gas pressures in gas ionization detectors vary according to their use. Ionization chambers may contain air at atmospheric pressure, or argon up to pressures of 20 atmospheres. Proportional counters are operated from about one-fifth of an atmosphere up to 70 atmospheres, the latter when it is desired to absorb all the energy of a particle, such as a conversion electron, in order to record spectral distributions. At 70 atmospheres of 90-per-cent argon and 10-per-cent methane, a typical proportional counter may require an operating potential of the order of 10,000 V. Applied voltages, at pressures normally used, range from 500 to 5000 V, increasing with increasing counter size, gas pressure and ionization potential of the gas filling. Geiger-Müller counters operate, typically with gas-filling pressures between about one-tenth of an atmosphere and one atmosphere and voltages between 500 and 2000 V.

Those readers who wish to study these matters in greater detail should refer to the excellent paper, that has been already mentioned, by D. R. Corson and R. W. Wilson, and to other references in the bibliography given at the end of this chapter.

Liquids and solids are the best absorbers for more penetrating radiation such as β particles and γ rays. Liquid-gas ionization chambers have been tried and liquid-gas bubble chambers are in common use. For the routine detection of more penetrating ionizing radiation at the present time, use is made of the fact that certain liquids and solids when exposed to such radiation act as transducers that emit light or, for a very short time, conduct electrons in the same manner as a gaseous ionization chamber. Such liquid- and solid-state detectors comprise the broad classes of scintillation and semiconductor detectors.

Scintillation and Semiconductor Detectors

Introduction. In addition to detectors of radiation based on gas ionization that have been discussed in the preceding sections, there is a broad group of solid-state, liquid, and gaseous detectors that respond to radiation either by the emission of light or by the conduction of electricity. Their classification is, however, overlapping and somewhat confused. Thus there are gases, chiefly the rare gases, whose atoms can be excited by radiation and which emit detectable radiation of visible light in the process of electron de-excitation. There are also organic crystals, such as anthracene and naphthalene, in which the electrons in the relatively loosely bound complex molecules can be excited by the interaction of incident radiation and then re-emit energy in the form of light, and can do so in the solid or vapor state, or in solution, or in a "solid" solution such as a plastic. It was also discovered by P. A. Čerenkov in 1934 that if very energetic particles traverse a transparent medium with speeds exceeding the velocity of light in that medium, visible radiation is emitted in a forward direction relative to the direction of motion of the particle. Čerenkov counters based on this principle find their chief application, however, in the physics associated with high-energy accelerators. The last and perhaps, at this time, the most important detectors are the crystalline solids such as sodium iodide, germanium, silicon, and zinc sulphide, which can be used as radiation detectors, either by the emission of light or as solid-state ionization chambers. These last consist of semiconductors, which have resistivities intermediate between those of conductors and insulators.

The inorganic crystals in the solid state and various complex organic scintillators in solution comprise the most widely used detectors of nuclear radiation. The loss of energy per unit path length (dT/dx in Chapter 3) is greater in a scintillator (as a liquid or a solid) or a solid-state ionization detector than in a gas detector, hence their application to the detection of high-energy radiations.

In all inorganic crystals, individual atoms are bound together in a lattice by forces between their outer-shell, or valence, electrons. In individual atoms the atomic electrons occupy well defined energy levels, but in a cluster of atoms in a crystal lattice the mutual interactions between the valence electrons cause their collective energy levels to merge into a wider-energy valence band which includes the levels of all valence electrons in the crystal. This valence band is depicted by the lower band shown in the energy-level diagram in Fig. 6-13. Valence

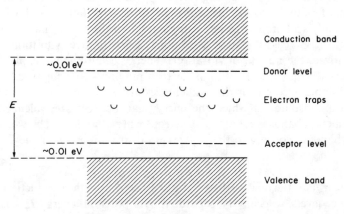

FIG. 6-13   Crystal band-gap energy-level diagram.

electrons having energies falling within the upper and lower limits of the valence band can move through the crystal by *exchanging* with valence electrons in adjoining atoms, but they are still bound to one or more atoms.

Electrical conduction in an inorganic crystal is envisaged to occur when the valence electrons receive additional energy by thermal or other processes and are removed from the valence band, in a manner analogous to ionization in a single atom; they now enter the conduction band, depicted in Fig. 6-13, in which they can move *freely*, apart from lattice interactions, throughout the crystal. The energy gap between the valence and conduction bands is E and, in a perfect crystal of a pure substance, this gap represents a forbidden region analogous to the forbidden regions between electronic shells in a single atom.

The valence band is normally filled with electrons that bond the atoms together in the crystal lattice. The conduction band may or may not contain free electrons, depending on the magnitude of the energy E separating the two bands and the temperature of the crystal.

In a conductor, such as graphite or a metal, the valence and conduction bands are contiguous or overlapping, and electrons can move freely from the valence band to the conduction band and will be directed as a current through the conductor upon the application of an electric field. In so doing, however, they will encounter obstacles in the form of bound atomic electrons in the crystal lattice and will be elastically scattered, thus imparting energy to the crystal lattice in the form of increased lattice vibrations. Such lattice vibrations are quantized, the quanta of vibrational energy being known as *phonons*. Differences in the energetics of electron-phonon scattering in different conductors account for their differences in resistance.

In a substance in which the energy gap between the valence and conduction bands is equal to E, electrons will be distributed between the two bands according to the Boltzmann distribution law, namely

$$n_c/n_v = e^{-E/kT} \qquad (6\text{-}12)$$

where n_c/n_v is the relative distribution of electrons in the conduction and valence levels, respectively, separated by a potential energy E, k is the Boltzmann constant and T is the absolute temperature. For T equal to 300K, kT is equal to 0.0259 eV, so that n_c/n_v will be equal to $1/e$ for $E = 0.0259$ eV, but will be far smaller for larger E.

The proportion of electrons in the conduction band will thus decrease as E gets larger, but increase with increasing temperature.

In a semiconductor the energy gap E between the valence and conduction bands may be of the order of 1 eV and, at room temperature, n_c/n_v, the ratio of free electrons in the conduction band to bound valence electrons, is of the order of 2×10^{-17}. As the temperature decreases this ratio becomes less and less, and in a pure semiconductor at the absolute zero all valence electrons would be confined to the valence band and a perfect crystal becomes a perfect insulator. By contrast, the resistance of a metal generally decreases with decreasing temperature. In general, the resistivity of a semiconductor is equal to $Re^{-w/kT}$ where R is the resistivity at infinite temperature, and w is an excitation energy. The values of this resistivity, R, for pure germanium and silicon are of the order of 10^{-4} ohm cm.

In an insulator the value of E is greater than a few electron-volts, which is very significantly greater than kT at ambient temperature, and which therefore results in correspondingly greater resistivities. In known substances resistivities at room temperature differ by a factor of more than 10^{25}. Thus good conductors such as silver and copper have resistivities at room temperature of 1.6×10^{-6} and 1.7×10^{-6} ohm cm respectively, while that of tellurium is 0.2 ohm cm. Good insulators have resistivities at room temperature of greater than 10^{16} ohm cm, examples being amber ($\sim 10^{16}$ ohm cm), sulphur (2×10^{17} ohm cm), sapphire (10^{16} to 10^{18} ohm cm) and fluorinated ethylene propylene (10^{19} ohm cm). Semiconductors have resistivities intermediate between those of metallic or graphite conductors on the one hand, and insulators on the other. As an example, the two semiconductors most commonly used as radiation detectors, namely germanium and silicon, have, in the pure state and at room temperature, respective resistivities of about 65 ohm cm and 2.0×10^5 ohm cm.

Even the purest crystals are, however, rarely free from small traces of impurities or of defects in their structure. Such impurities and defects create absorption and emission energy levels, and "electronic traps" intermediate between the energy levels of the valence and conduction bands. Such extra energy levels are illustrated in Fig. 6-13. Their presence greatly modifies the luminescent and electrical properties of a crystal. Most *electron traps* are situated at energy levels a few hundredths of an electron-volt or more below the conduction band. Electrons that have been excited into the conduction band will generally migrate around the crystal and some will de-excite to the valence band, emitting radiation, while others will interact with a trapping defect or impurity, losing energy and being trapped at an energy level from which transitions to the valence band do not necessarily occur. If the energies of the trapping levels are not too far below that of the conduction band, trapped electrons can again be raised to the conduction band by thermal excitation from the surrounding lattice, thence returning to the valence band with the emission of delayed luminescence. At ambient temperatures, and in the absence of exciting radiation, metastable trapping levels that are a few hundredths of an electron-volt below the conduction band are rapidly emptied by thermal excitation of the electrons to the conduction band, whence they can now return to the valence band. Trapping levels at larger energy differences below the conduction band can only be emptied to the conduction band with the use of higher temperatures, and thermoluminescent dosimeters (TLD's) are based on this principle.

Although there seems to be, as yet, no consistent use of the terms "fluorescence" and "phosphorescence," prompt radioluminescence is usually called *fluorescence* and radioluminescence that is delayed for than about 1 μs after removal of the source of exciting radiation is usually referred to as *phosphorescence*. "X-ray fluorescence" and "fluorescence yield" are terms that exemplify prompt luminescence.

In the case of crystals used as semiconductor radiation detectors, or so-called solid-state ionization devices, not only do the electrons that have been elevated to the conduction band move under the influence of an externally applied electric field, but the vacancies in the lattice, or "holes," from which these electrons have come in the valence band, can act as conductors of electricity by the process of acquiring a valence electron from, and creating a "hole" in, a neighbouring atom which can then, in turn, acquire an electron from its neighbour, and so on. The externally applied electric field thus also causes a flow of electrons in the valence band, or, as it is often called, a flow of positive holes in the opposite direction.

Such, in rather brief outline, are the principles underlying the properties of solid-state scintillation and semiconductor radiation detectors. More detailed descriptions are given, for example, by Birks (1964) and in various articles in the *Encyclopaedic Dictionary of Physics* (1962). The operation of crystal and liquid scintillators, and of solid-state semiconductor ionization devices will be discussed in the following sections of this Chapter.

Scintillation Detectors. One of the earliest instruments used for the detection of radiation was the spinthariscope, devised by W. Crookes in 1903, in which α particles bombarding a screen of activated zinc sulphide produced light flashes of sufficient intensity that they could be observed through a low-power microscope in a darkened room. This property was also observed by J. Elster and H. Geitel in the same year. The process whereby the energy deposited by the incident ionizing radiation in the scintillator is converted into light, is the basis for modern methods of scintillation counting. Various refinements, such as the substitution, by S. C. Curran and W. R. Baker in 1944, of an electron-multiplier phototube for the microscope and human eye, followed by the development of phosphors other than activated zinc sulphide, have significantly increased the utility of present-day scintillation detectors. Such detectors may now be applied to the detection, and measurement of the energy distribution, of high-energy α and β particles, and of photons with energies greater than approximately 2 keV.

While activated zinc sulphide is still used for α-particle detection, numerous phosphors, both organic and inorganic, have been discovered and developed since the late 1940's. The mechanisms responsible for producing scintillations in organic and inorganic phosphors differ. Also, organic scintillators are usually restricted in their use to charged-particle or very low-energy-photon measurements because of their low atomic number, Z, and density, whereas inorganic-crystal scintillators, with their higher Z and density, are most commonly applied to the detection of x and γ rays with energies greater than 2 or 3 keV.

Organic- and Inorganic-Crystal Scintillators. Following Crookes' discovery in 1903 that activated zinc sulphide scintillated in response to α-particle bombardment, the next significant step in the development of solid-state scintillators did not occur until 1947 and 1948 when H. Kallmann and then P. R. Bell used single crystals of naphthalene and anthracene, respectively, for the detection of ionizing radiation. Anthracene has been used extensively for the detection of β particles, and is about 700 times as dense as, say, argon at atmospheric pressure, as used in a proportional counter. Its molecules de-excite by the emission of photons with wavelengths around 4450 Å (445 nm), and decay times of about 30 ns.

Also in 1948, R. Hofstadter discovered the scintillating properties of thallium-activated sodium iodide crystals. Their density of some three times greater than that of anthracene meant that they were that much more desirable for the detection of γ rays.

Sodium iodide and other alkaline halide crystals are perhaps the best known class of inorganic crystalline scintillators, and thallium-activated sodium iodide, NaI(Tl), crystals comprise the most widely used type of scintillation photon detector.

Although pure NaI crystals are very efficient scintillators at 77 K, with very fast response, they are some ten times less efficient at room temperature. As we have seen, the electrical resistivity in the conduction band is an exponential function of w/kT, and the mobilities of electrons in the conduction band depend on complicated crystal-lattice scattering functions of the temperature. The radioluminescence-efficiency dependence on temperature must therefore be rather complex, but it is clear from experiment that, at ambient temperatures, a greater proportion of the radiant energy deposited in the crystal is dissipated in radiationless transitions such as crystal vibrations (phonons), or heat. Even when operated at their optimum temperature, for maximum scintillation efficiency, pure NaI crystals are limited to a useful thickness of about 1.3 cm because of the strong self-absorption of their fluorescence radiation at a wavelength of 3000 Å (300 nm).

The introduction of thallium into the NaI melt at a concentration of 10^{-3} mole fraction or 0.2% Tl by weight, prior to growing the crystal, creates energy levels in the crystal lattice below the lower limit of the conduction band. The thallium impurity forms halide complexes known *luminescence centres* in the crystal lattice. Thus, for example, an energetic photon interacting with the crystal can generate an energetic Compton recoil electron which, through inelastic collisions in the crystal lattice, can raise electrons into the conduction band. These electrons, in their progress around the crystal lattice, can then transfer energy to many luminescence centres, which, in turn, de-excite by radiative transitions in a fairly broad band of wavelengths with a maximum intensity at about 4300 Å (430 nm). Unlike NaI crystals, NaI(Tl) crystals are transparent to their own luminescence and their usable thickness is therefore not restricted. With the addition of increasing amounts of thallium, the intensity of the 300-nm NaI emission band decreases as that of the 430-nm NaI(Tl) emission band increases. This indicates a transfer of energy from NaI luminescence centers to the thallium. It should also be noted that the gamma-ray-scintillator efficiency of NaI(Tl) crystal at 300 K is intermediate between those of pure NaI at 77 K and 300 K (see Birks, 1964, for further references).

Crystal-Scintillation Detector Systems. The operation of a typical organic- or inorganic-crystal scintillation-detector system can be broken down into a series of five sequential processes, each with an associated efficiency. These processes, in their order of occurrence, are:

1. Deposition of energy by the incident ionizing radiation in the crystal, producing excitation and ionization in it.

2. The emission of light in the processes of de-excitation.

3. Collection of this light on the photosensitive cathode of the electron-multiplier phototube.

4. Emission of photoelectrons from the photocathode of the phototube.

5. "Multiplication" of the number of electrons emitted from the photocathode in the dynode system of the phototube.

These processes, operating in sequence over a very short time interval (of the order of a microsecond or less), produce a current pulse at the anode of the electron-multiplier phototube. The amplitude of this current pulse is proportional to the energy deposited in the scintillator by an individual event.

The first process is illustrated in Fig. 6-14 showing the interactions of three gamma rays of different energies with a NaI(Tl) crystal. In this example, the lowest-energy gamma ray (γ_1) is shown producing an electron (e_p) by a photoelectric interaction with an extranuclear electron of one of the Na, I, or Tl atoms followed by x-ray emission from the excited atom. The intermediate-energy gamma ray (γ_2) is shown interacting with the crystal through multiple Compton scatterings of the gamma ray, producing Compton electrons (e_c), and culminating in a final photoelectric interaction. The highest-energy gamma ray (γ_3), with energy in excess of 1.022 MeV, first produces a positron-electron pair (e^+, e^-) in the Coulomb field of the nucleus of an atom of the crystal. The positron then annihilates, producing two 511-keV photons (m_0c^2), themselves interacting with the crystal in a manner analogous to the interactions of γ_1 and γ_2 (see Chapter 3).

The Compton recoil electrons and photoelectrons, most of which have energies well above that of the conduction band, now travel around the crystal transferring energy in inelastic collisions to atoms in the lattice, exciting these atoms or raising their valence electrons into the conduction band. The whole lattice then de-excites with the emission of a band of luminescent radiation with a peak at about 430 nm.

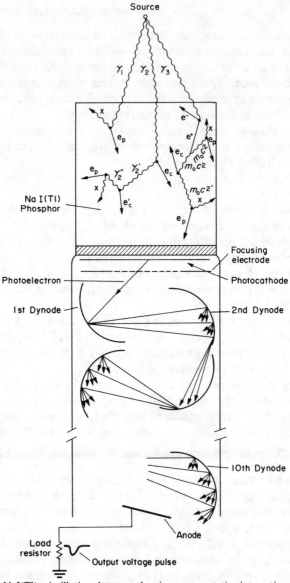

FIG. 6-14 NaI(Tl) scintillation detector, showing representative interactions of three γ rays of differing energies with the phosphor. For clarity the ultraviolet photons produced by these interactions are not shown.

As we have seen, not all of the radiation energy deposited in a crystal scintillator such as NaI(Tl) is converted into energy of luminescence, the major part of the deposited energy being dissipated in other complex processes such as lattice vibrations, or heat. In fact, in the case of NaI(Tl) at ambient temperatures, only about 10 per cent of the radiation energy absorbed in the crystal is re-emitted as luminescent energy. The ratio of luminescent to absorbed energy is known as the *energy conversion efficiency* or the *intrinsic efficiency* of the scintillator, with typical values for different scintillators ranging from a fraction of 1 per cent to around 40 per cent.

Considering the case of NaI(Tl) at ambient temperatures we see that the deposition of 3000 eV of radiation energy in the crystal will give an output of around 300 eV of luminescent energy. The wavelength of this luminescence, of about 430 nm, corresponds to photons of 3 eV, so that the luminescence output will comprise approximately 100 photons.

Before this light can, however, be converted to an output signal of electrons from the phototube, efficiencies of two other intermediate steps must be considered, namely the *optical efficiency* of collecting the luminescence photons on to the photocathode of the phototube, and the *photoelectric efficiency* of the cathode in converting incident luminescence photons into photoelectrons.

To optimize the overall efficiency of a crystal-scintillation detector, it is important to provide for the best possible light collection by the cathode of the phototube. In the first place, it is essential to select a crystal-activator combination that is highly transparent to its own luminescent radiation. The face of the phototube is then usually coupled optically to a flat surface of the crystal using a jelly with suitable refractive index, and the other surfaces of the crystal are surrounded by a good reflecting material such as aluminum oxide. In the case of hygroscopic material such as NaI, the whole is then hermetically sealed into a thin aluminum enclosure, that is sometimes provided with a thin beryllium window. With the taking of such precautions, optical efficiencies, namely the fraction of photons produced in the scintillator that reaches the photocathode, can be as high as 0.9.

As the luminescence wavelengths of different scintillators vary it is, of course, important to match spectral sensitivity and the envelope material of the phototube to the emitted luminescence. For example, glass becomes opaque to radiation in the ultraviolet and it may be necessary to use a phototube that has a window of quartz.

The photoelectric efficiency of the photocathode is a function of the photoelectric and optical absorption characteristics of the cathode material and the wavelength of the incident photons. Values range from 0.05 to 0.10 photoelectrons per incident photon.

Thus we see that for 3000 eV of radiation energy deposited in an NaI(Tl) crystal having an intrinsic efficiency of 10 per cent, the luminescence output will consist of about 100 3-eV photons. An optical efficiency of 90 per cent will reduce the number of photons to about 90 incident on the photocathode which will emit 9 photoelectrons if the photoelectric efficiency is 10 per cent. In this example, we see that for 3000 eV of radiation energy deposited in the crystal, 9 photoelectrons will be emitted from the cathode of the phototube (i.e. one photoelectron from the phototube cathode for every 0.3 keV of radiation energy deposited in the crystal).

As these photoelectrons leave the cathode, they are directed by the electrostatic field to the first multiplier electrode, or *dynode*. This electrode has the property of emitting three, four, or five electrons for every single electron which strikes its surface with an energy of, say, 75 to 150 eV (hence the designation "electron multiplier"). There may be from 10 to 14 such multiplier stages in a given tube, each of which is maintained 75 to 150 V more positive than its predecessor. Thus, from the emission of *one single electron* from the cathode of the phototube, a pulse of from *one million* to *a hundred million electrons* may impinge upon the final stage in the tube, namely its anode. The anode is usually connected to a load resistor external to the tube; the pulse of electrons, flowing through this resistor, produces a negative-voltage pulse which is then amplified and analyzed by suitable electronic equipment.

When a scintillator is excited by a single ionizing event, the intensity of the de-exciting pulse of radioluminescence rises rapidly to a maximum that is proportional to the amount of energy deposited in the scintillator. The rise to maximum intensity occurs in about 10^{-9} s and then decays exponentially with time. The time required for the light intensity to decrease to $1/e$, or to approximately 37 per cent, of the maximum is referred to as the *decay time constant* of the scintillator and is usually designated as τ. This time constant is a specific property of each scintillator, organic scintillators being characterized by decay time constants of 10^{-8} s, or less, and those of inorganic crystal scintillators being typically of the order of 10^{-6} s.

Liquid-Scintillation Counting. Another type of scintillation counting is that in which the radioactive sample is dissolved in an organic solvent together with one or more organic *fluors*. This method of liquid-scintillation counting is used mainly for the detection and assay of radionuclides that decay by α-particle and β-particle emission, and especially for relative measurements of low-energy β-particle emitters such as ^{3}H and ^{14}C. It has also been applied to measurements of relative activities of radionuclides that decay by electron capture followed by the prompt emission of x rays and Auger electrons.

A liquid scintillator absorbs energy from radiation by the ionization and excitation of the solvent molecules. This energy is then efficiently transferred to the organic-fluor molecules, the concentrations of which usually range from 0.5 to 1 per cent. A fraction of this energy is then converted into radioluminescence by de-excitation of the fluor molecules. The efficiency for the conversion of radiation energy deposited in the scintillator to that emitted in the form of radioluminescence, analogous to the energy-conversion efficiency of inorganic crystals, is usually of the order of 3 per cent. The wavelengths of this fluorescence correspond to those of photons in the violet or ultraviolet with energies, hv, of about 3 eV.

A second fluor, often called a *wavelength shifter*, may be incorporated into the scintillator solution in order to match the spectral response of the phototube to that of the radioluminescence. The concentration of the secondary fluor, when used, is of the order of 10 or 20 per cent by mass of that of the primary fluor. However, many primary scintillators now in use have spectral matching factors (the degree of overlap of fluorescence wavelengths and phototube spectral response) of over 95 per cent.

Solvents commonly used are toluene, xylene, dioxane, and other aromatic hydrocarbons or ethers. Dioxane is important because it is a solvent not only for organic fluors but also for water and aqueous solutions. Typical fluors are *p*-terphenyl (TP) and 2,5-diphenyloxazole (PPO). 1,4-*bis*-(2-(5-phenyl-oxazolyl))-benzene (POPOP) is often used as a wavelength shifter. Plastic scintillators consist of solid solutions of organic fluors in transparent plastics.

In liquid scintillators, the efficiency of conversion of deposited radiation energy into radioluminescence can be greatly decreased by the presence of substances that dilute and decrease its effectiveness in absorbing radiation energy, or by the addition of chemicals that de-excite the solvent molecules without the emission of photons, or by others such as colouring materials that act as filters in absorbing the fluorescence photons. These three processes are referred to collectively as *quenching*, and *quenchers* or *quenching agents* should be avoided. Oxygen is a common quenching agent that can be removed from liquid scintillators by bubbling an inert gas, such as argon or nitrogen, through the solution.

As with an organic or inorganic crystal scintillator, the overall efficiency of a liquid-scintillation system is a function of the energy-conversion efficiency of the scintillator, the optical efficiency of fluorescence-photon collection, and the photoelectric efficiency of the phototube cathode in converting fluorescence photons to photoelectrons. A typical overall efficiency in terms of radiation energy deposited in the liquid scintillator per photoelectron produced from the phototube cathode is 2 keV per photoelectron. This compares with the order of 0.3 keV per photoelectron for a NaI(Tl) crystal.

A more important specification of any scintillation-detector system is the overall *counting efficiency*, namely the fraction of output pulses from the phototube per decay of a radionuclide incorporated in the scintillator. For a good commercial liquid-scintillation system the counting efficiency may vary from about 60 per cent for ^{3}H to nearly 100 per cent for higher-energy beta emitters.

The output signal from the phototube is proportional to the radiation energy deposited in the liquid scintillator. When liquid-scintillation-counting systems are used for the assay of low-energy beta-emitting nuclides it is desirable to reduce "noise" output pulses from the phototube, arising from the thermal emission of electrons from the photocathode, in order to observe the pulses corresponding to energies as far as possible into the low-energy end of the beta-particle spectrum. Reduction of background phototube noise can be effected, however, by using two phototubes to collect the fluorescence photons from the scintillator, with electronic circuitry that records only those output pulses from the phototubes that are coincident within a short interval of time. The recorded number of background noise pulses, that are uncorrelated in time between the two phototubes, will thereby be significantly reduced, whereas fluorescence pulses that are collected in coincidence will be recorded. The photoelectric conversion efficiency will be decreased but a 2-keV to 3-keV electron, if it deposits all its energy in the scintillator, should still give a measurable pulse from each phototube. The phototube backgrounds can be further reduced by cooling them in a refrigerating unit, thereby reducing the thermal emission of electrons from the photocathode. Specially selected low-noise phototubes, for use at room temperatures, are now also becoming rather widely available.

For a more comprehensive treatment of the theory of radioluminescence and of the operation of scintillation detectors, those interested should consult the *Encyclopaedic Dictionary of Physics* (1962), Birks (1964), Price (1964), Adams and Dams (1970), NCRP (1978), and references given therein.

Semiconductor Detectors. We have seen in the introduction to this section on Scintillation and Semiconductor Detectors that electrons in the valence band of an inorganic crystal can be elevated to the conduction band by the interaction of radiation with the crystal. This can also occur by the scattering of phonons (lattice vibrations) by the valence electrons. The resistivities of crystals therefore depend on the energy gap, E, between the valence and conduction bands and the temperature. As we go from crystalline conductors to insulators, E increases from zero for graphite to about 6 eV for diamond. Diamond is almost an insulator having a resistivity of the order of 10^8 ohm cm, but both diamond and substances such as pure crystals of germanium and silicon, which at room temperature have resistivities of about 65 ohm cm and 2×10^5 ohm cm, respectively, are classed as semiconductors. On the other hand, as the temperature is reduced, kT decreases and all pure defect-free semiconductors at the absolute zero of temperature are perfect insulators.

In the case of diamond it is clear that, if there were no dissipation of energy by other processes, the interaction of 6 eV of ionizing radiation with the crystal would create one electron-hole pair, compared with

15.7 eV required to create one electron-ion pair in the case of argon (i.e. the ionization potential of argon). With its much greater density diamond would, in theory, make a very attractive solid-state ionization chamber, compared with a pulse ionization chamber containing argon. Unfortunately, however, diamond cannot be obtained in a pure state and its semiconductor properties are due to natural impurities and lattice imperfections, that also cause it to have a superabundance of traps. Thus, while it was found to perform as expected as a detector of ionizing radiation, it was also found to build up excessive space charge due to trapped electrons which, in turn, resulted in a decrease in amplitude of the output current pulse with increasing numbers of detected ionizing events. It was therefore necessary, after a period of use, to remove this space charge by annealing the crystal or by reversing the applied electric field.

Attention turned, therefore, to the development of other semiconductor materials, and, at the present time, the most widely used are germanium and silicon crystals of high quality and purity, to which, usually, controlled quantities of selected trace impurities have been added. Other semiconducting materials, notably gallium arsenide and cadmium telluride, have also been investigated.

It has been stated, and illustrated in Fig. 6-13, that, because of crystalline imperfections and the presence of impurities, electron-absorption and electron-emission levels, and electron traps, are created at energy levels between the valence and conduction bands. These absorption and emission levels are created by the so-called *acceptor* and *donor* impurity atoms, respectively, and these can also be introduced in controlled quantities into the "melt" from which the semiconducting crystals, such as germanium and silicon, are grown. Impurity atoms having valences different from that of the host crystal create *valence defects*, that create sites in the crystal that can provide electrons (donors) or accept electrons (acceptors). In crystals of compound semiconductors such as gallium arsenide, these defects can also arise from non-stoichiometry. In elemental semiconductors the bonding between atoms is *covalent*, in that two electrons with anti-parallel spins are shared between two atoms in the manner of the electron-pair bond of chemical compounds. In compound crystals, such as sodium chloride, the bonding is ionic, and arises from the Coulomb attraction between positively and negatively charged ions. In some compound semiconductors such as zinc sulphide the bonding may also be covalent.

The crystal structures of germanium and silicon are face-centred cubic like that of diamond, with covalent tetrahedral bonding. Diagrammatically this can be depicted as in Fig. 6-15 which represents,

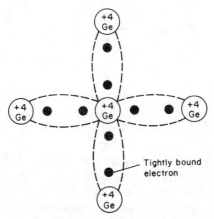

FIG. 6-15   Crystal-lattice model of pure germanium.

in two dimensions, one "building block" of a three-dimensional structure, the dashed lines linking pairs of atoms and representing the electron-pair or covalent bonds.

If, however, a trace quantity of pentavalent atoms such as phosphorus or antimony is introduced into the crystal structure, the atoms of these "impurities" will displace germanium or silicon atoms in their respective lattices in the manner shown in Fig. 6-16. Each of these impurity atoms will contribute an extra loosely-bound electron which can be readily elevated to the conduction band by the thermal energy available at room temperature, kT being of the order of 0.026 eV. The effect of these donor atoms is to increase the conductivity of the germanium or silicon. This type of semiconductor is known as *n-type* because the charge carriers are negative electrons. The controlled introduction of impurity atoms is called *doping*.

If, on the other hand, germanium, say, is doped with trivalent atoms such as aluminum or gallium, acceptor sites are created. Electrons can

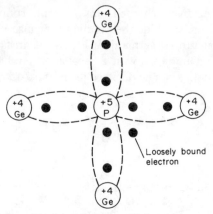

FIG. 6-16   Crystal-lattice model of germanium with n-type impurity.

now be readily elevated from the valence band of the crystal to the acceptor level by means of the thermal energy available in the lattice, and the vacancies, or holes, left in the valence band act as positive charge carriers. Such a semiconductor is therefore known as *p-type*, because the vacancies, or holes, left in the valence band act as positive charge carriers, moving to the cathode under the influence of the applied electric field. It is convenient to visualize the motion of a hole in the valence band according to the semi-classical picture of the hole moving to the negative collecting electrode by virtue of numbers of electrons moving one interatomic distance in the opposite direction. R. G. Hibberd has likened this to one vacant theatre seat, in the middle of a row of seats, "moving" to the aisle, because the intervening patrons all move over one place away from the aisle. However, in the quantum-mechanical description of a crystal lattice, the physical behaviour of a hole is described in terms of the totality of electrons in the band and not in terms of the motions of single electrons. The behaviour is analogous to that of a bubble moving up through a column of water; the motion of the bubble, or hole, can be readily described, but it would be both futile and meaningless to attempt to describe its motion in terms of the motions of the individual water molecules.

A hole moving in a nearly filled valence band is analogous to, and can be considered to be just as "real," as a positron which, according to

Dirac's theory, is a hole moving in a nearly filled "sea" of electrons in negative-energy states (see Chapter 3).

Such p-type doping is illustrated in Fig. 6-17, and it should be remarked that most crystals of germanium and silicon, by virtue of their natural impurities, are, in fact, p-type.

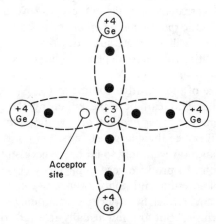

FIG. 6-17   Crystal-lattice model of germanium with p-type impurity.

The dominant charge carriers, namely holes in p-type material and electrons in n-type material, are known as *majority carriers*. The *minority carriers* are electrons in p-type material and holes in n-type material.

One of the chief problems to be overcome in the use of semiconductors as solid-state ionization chambers lay, however as their name implies, in the fact that they are not good insulators and that the leakage current resulting from the application of an electric field, intense enough to sweep the electrons and holes swiftly to the collecting electrodes, is too large. As in a gas ionization chamber, the background leakage current must be minimal for optimum signal-to-noise ratios. In the case of germanium and silicon crystals, most of which tend to be p-type, this problem has been approached by one fundamental method, namely the removal of charge carriers from the bulk material. Three different procedures have so far been developed to achieve this end, but, in all of these, the flow of current through the detector is inhibited by the use of rectifying contacts.

The first procedure is to create the so-called p-n-junction rectifying diode. This is achieved by diffusing high concentrations of donor atoms N_D into p-type material, or of acceptor atoms N_A into n-type material. At the p-n junction, or interface, between the p- and n-type regions, a space-charge potential is created that depletes the region of charge carriers. This region is known as the *space-charge region* or the *depletion region*, and its mode of operation will be more fully described in the next section. However, the sensitive volume, or depletion region, of such detectors has a depth that is proportional to $(V/|N_A - N_D|)^{1/2}$, and, therefore, in order to maximize the size of the sensitive volume $|N_A - N_D|$ must be made as small as possible. This can be done either by the second or third procedures, namely by *purification* or by the *compensation* of the impurities that produce the charge carriers.

Purification has, however, only so far been successfully applied in the case of germanium to produce very high-purity crystals in which the concentration of acceptor and donor atoms, $N_A + N_D$, is not greater than about $10^{10} \, cm^{-3}$. Such high-purity germanium crystals were at one time called "intrinsic," but are now usually referred to simply as *high-purity*, the term *intrinsic* being applied to "compensated" germanium or silicon material in which the difference in the concentrations of uncompensated charge carriers, $|N_A + N_D|$, has been reduced to the order of $10^{10} \, cm^{-3}$, or less.

Compensation is achieved by drifting ions, generally lithium ions, that are themselves electron donors interstitially into p-type silicon or germanium crystals. Here they form complexes with the acceptor impurities, thereby neutralizing the acceptor atoms in their rôle as creators of positive charge carriers (holes). Because of the high mobilities of lithium ions in germanium, quite large lithium-compensated germanium, Ge(Li), detectors can be made by this method. Such compensated Ge(Li) detectors having sensitive volumes greater than $100 \, cm^3$ can now be produced, and the high atomic number of germanium makes it a most suitable material for the total absorption of high-energy photons.

In addition, the conductivities of detectors produced by each of the above procedures can be reduced still further, when required in practice, by cooling them to the temperature, say, of liquid nitrogen (77K).

In a good detector, it is also important to minimize electron trapping into delayed metastable levels, in order to avoid the delayed collection of the charge carriers or the build up of space charge. This can normally be achieved by the preparation and use, in the above procedures, of very pure crystals that are as free as possible from crystal defects.

In high-purity or lithium-drifted intrinsic germanium or silicon detectors, the absolute number of excess positive or negative charge carriers, $|N_A - N_D|$, is of the order of 10^9 to 10^{10} cm^{-3}, or less. Those semiconductors, in which the electrical conductivities have been greatly increased by the uncompensated presence of impurities are known as *extrinsic* or *impurity* semiconductors. These latter can be p-type or n-type, produced usually by controlled doping. Semiconductor materials that have been heavily doped are often designated as p$^+$-type or n$^+$-type, while high-purity germanium which is still not quite intrinsic is often called π (very slightly p) or ν (very slightly n).

The mobilities of electrons in the conduction band are respectively 1800, 3800 and 1600 cm^2 V^{-1} s^{-1} for diamond, germanium and silicon; the mobilities of holes in the valence band at 300K and electric field strengths of 1000 V cm^{-1} for the same materials are respectively 1200, 1800 and 400 cm^2 $V^{-1} s^{-1}$. Thus with an applied electric field of, say, 250 V cm^{-1}, drift velocities of the negative and positive charge carriers will be of the order of 10^6 cm s^{-1} and 5×10^5 cm s^{-1} in germanium. High-purity germanium has electron and hole mobilities, respectively, of 3.6×10^4 and 4.2×10^4 cm^2 $V^{-1} s^{-1}$ at 77 K and the corresponding values for compensated material are 10^4 and 2×10^4 cm^2 V^{-1} s^{-1} at 77 K. Thus in a Ge(Li) detector at 77 K, drift velocities will be of the order of 10^6 cm s^{-1} for electric fields of the order of 500 to 1000 V cm^{-1}, and collection times for a depletion layer of 20 mm would be of the order of 100 ns. Drift velocities of electrons and positive ions under typical operating conditions in a proportional counter were, respectively, of the order of 10^6 cm s^{-1} and 10^3 cm s^{-1}. It is therefore clear that collection times of both the negative and positive charge carriers in a semiconductor detector will, in general, be much shorter than the electron : collection time in a proportional counter, in which the collection time for the positive ions is also much longer. The shape of the output pulse from a semiconductor detector is therefore, for all practical purposes, not dependent on the location of the ionization within the detector, as it is, however, in the case of a proportional counter or pulse ionization chamber.

The p-n Junction. Consider a thin crystal of lightly doped p-type silicon into one surface of which phosphorus has been diffused, by heating, to form a heavily doped n^+ region having a depth of about 10^{-3} mm. Thin films of aluminum or gold are evaporated on to the opposite faces to facilitate the application of an electric field and the remaining surfaces are normally etched to reduce residual surface electrical conductivity. Such a device, known as a p-n^+ or p-n junction, is illustrated diagrammatically in Fig. 6-18.

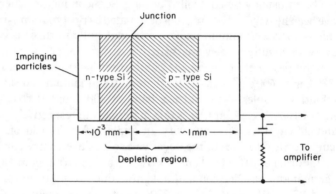

FIG. 6-18   Cross-sectional drawing of *p-n* junction semiconductor detector.

To understand the operation of such a p-n junction, let us also consider its energy level diagram as shown in Fig. 6-19.

This junction is formed from *one crystal*, in the n^+ region of which the donor atoms will be contributing thermally excited electrons to the conduction band (i.e. transitions such as A in Fig. 6-19). If the n^+ region had been an isolated region it would have contained equal numbers of conduction electrons and positively charged donor atoms, thus maintaining the electrical neutrality of that region. However, in the p region of the same crystal, electrons will have been elevated to acceptor

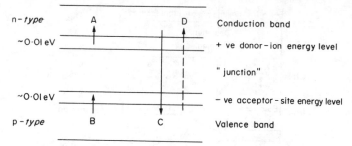

FIG. 6-19   Semiconductor energy level diagram illustrating transitions A, B, C and D
referred to in the text.

sites (transition B in Fig. 6-19 and again, if this region had been isolated, there would have been equal numbers of negatively charged acceptor atoms and positive charge carriers, in the form of holes in the valence band.

But we do not have two isolated regions but an n^+ region and a p region *in one and the same crystal.* Moreover, the electrons that have been raised to the conduction band in the n^+ region are free to move *throughout* the *whole crystal.* The majority carriers (electrons) generated in the n^+ region are therefore free to combine with the majority carriers (holes) generated in the p region (transition C in Fig. 6-19). Every such combination of an electron and a hole will leave one uncompensated positively ionized donor atom in the n^+ region and one uncompensated negatively ionized acceptor atom in the p region, and many such combinations will give rise to the same equal numbers of ionized donor and acceptor atoms in each region (Fig. 6-20). Donor and acceptor sites should be distributed fairly uniformly throughout each region of the crystal, as illustrated in Fig. 6-20b. The majority carriers in each region, namely electrons in the n^+ region and "holes" in the p region, are also shown in Fig. 6-20b. In this simple illustration, minority carriers are not shown as their numbers would be many orders of magnitude less than the majority carriers in each region.

In the process of fabrication, as soon as the phosphorus donor atoms begin to diffuse into the p-type crystal, electrons in the conduction band will cross the junction and combine with holes in the p region just across

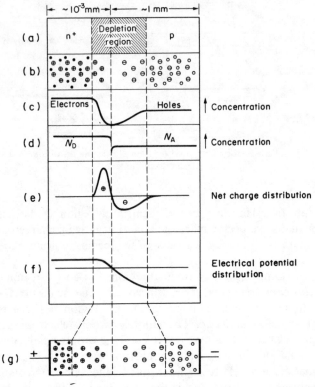

FIG. 6-20   Diagram of a p–n junction showing, non-quantitatively: (a) depletion region (bounded by dashed lines); (b) distribution of acceptors ⊖, donors ⊕, electrons ●, and holes ○ in an unbiased p-n junction; (c) electron and hole concentrations; (d) donor and acceptor concentrations; (e) net charge distribution (producing a dipole layer at the junction); (f) electrostatic potential distribution corresponding to the dipole layer; (g) distribution of acceptors, donors, electrons and holes in a reverse-biased p-n junction.

the junction, and thus give rise to uncompensated positive donor ions on one side of the junction and uncompensated negative acceptor ions on the other, as illustrated in Fig. 6-20b. Thus on each side of the barrier a space charge of opposite sign is established which results in an electrical potential difference, illustrated in Fig. 6-20f, that opposes the movement of any more electrons across the junction. The distribution of charge in this space-charge region on either side of the junction is illustrated in Fig. 6-20e. In the n^+ and p regions beyond the space-charge region, the ionized donor and acceptor atoms are electrically compensated, respectively, by the free electrons and free holes. These regions are, however, conducting and will be equipotential regions, the n^+ being positive with respect to the p region. Some of the electrons in the n^+ region will have enough energy to cross the potential barrier and these will combine with holes in the p region or be scattered and accelerated back into the n^+ region. Other electron-hole pairs will continue to be created thermally in the space-charge region (e.g. transition D in Fig. 6-19) and these will be accelerated back by the potential difference into the n^+ region and p regions, respectively. At equilibrium, however, the electron current in one direction will balance that in the opposite direction. Also, because the concentration of donors in the n^+ region is much greater than that of the acceptors in the p region, the space charge will extend much further into the p region than into the n^+ region.

It is clear that free electrons and holes cannot remain in the space-charge region. If an electron-hole pair is created in this region by, for example, a radiative interaction, the electron is accelerated to the n^+ side of the junction and the hole to the p side (i.e. electrons move in from the p region). The space-charge region that is depleted of charge carriers is called the *depletion region*, and the junction is usually referred to as a *p-n junction*.

The p-n junction also acts as a rectifying junction. Let us suppose that we connect such a junction to an external circuit, as in Fig. 6-18, but with a variable source of potential. First consider the case where the negative potential is applied to the n contact and positive to the p contact. Initially, for a value of the applied potential that is lower than that due to that induced by the space charge at the junction, no current will flow. As, however, the applied potential is increased, the space-charge field will be overcome, electrons in the conduction band will be driven across the junction, and a current will flow. The externally applied field in this direction is usually referred to as a *forward bias*. In terms of Fig. 6-18 electrons move into the n depletion region and out of the p depletion region with increasing applied voltage, and the Maxwell–Boltzmann equilibrium (Eq. 6-12) shifts to lower concentrations of positive-donor and negative-acceptor atoms. In the language of semiconductor physics, electrons from the n region and holes from the p region are driven towards the junction, thereby neutralizing the ionized donor and acceptor atoms, respectively, in the depletion region.

If, however, the externally applied electric potential is reversed (i.e. positive to n and negative to p), it will now supplement the potential due to the induced space charge. A potential applied in this direction is usually referred to as *reverse bias*. Conduction electrons will now be driven from right to left (Fig. 6-18), out of the n depletion region and into the p depletion region. As a result the Maxwell–Boltzmann equilibrium concentrations would tend to shift to more positively ionized donor atoms in the n depletion region and more negatively ionized acceptor atoms in the p depletion region. The total depth of the depletion region is therefore increased (Fig. 6-20g), and no current flows until the potential is increased beyond a break-down point and the current rapidly increases. In the language of semiconductor physics, electrons diffuse towards the positive terminal and holes towards the negative terminal thereby increasing the depth of the depletion region.

In the depletion layer under the reverse bias the valence band and acceptor sites must be completely filled with electrons, and the donor sites must be all ionized and the conduction band completely empty of electrons. In this state the depletion layer is equivalent to a very good insulator. If this were not the case, then transitions A or B (Fig. 6-19) could still occur and the electrons elevated to the conduction band or holes created in the valence band would be swiftly swept away by the reverse-bias field, thus completing the process of removing all charge carriers from the valence and conduction bands.

Therefore the only charge carriers that can be created in the depletion layer must be through a process such as D (Fig. 6-19). In the case of silicon, with its relatively large band-gap energy of 1.1 eV, the consequent bulk leakage current (i.e. through the body of the material) in a reverse-biased p-n-junction detector is only a few nanoamperes at 25°C with an applied voltage of about 100 V. This permits operation of a silicon radiation detector at room temperature. Germanium detectors, however, with their smaller band-gap energy of about 0.7 eV must be operated at a reduced temperature (usually at about 77 K, the boiling point of liquid nitrogen) in order to reduce the bulk leakage current to an acceptable value.

Surface leakage current is dependent on the physical form of the semiconductor crystal surface, surface contamination, and ambient temperature, and is more difficult to control. Surface contamination may, in fact, produce subsidiary p- or n-type conducting layers. Surface leakage current must be reduced by techniques such as etching the surface and by care in mounting the crystal.

Because the p-n junctions function as rectifiers, thus providing the high resistivity required for operation as radiation detectors, they are often referred to as *diodes*. Such junctions can be equally well prepared by diffusing a p^+ layer into lightly doped n-type material.

Ohmic Contacts. In the last section, in describing the preparation of a p-n junction, we said that thin films of aluminum or gold were evaporated on to the opposite faces to facilitate the application of an electric field, as illustrated in Fig. 6-18. It is, however, extremely difficult to make a satisfactory connection to a semiconducting material without introducing some kind of rectifying potential barrier. If the aluminum or gold were to diffuse into the semiconducting material there is always a possibility that a rectifying junction could be established. In an ohmic contact the current should be linear with applied voltage in both directions. Such conditions can sometimes be established by heavily doping the surface of the extrinsic material, as n^+ to n, or p^+ to p, and then evaporating a thin metal contact to the heavily doped layer. As illustrated in Fig. 6-18 the incident radiation, especially if corpuscular, should traverse only as thin a layer as possible before entering the depletion region. To some extent, the fabrication of a good ohmic contact is more an art than a science.

Surface-barrier Detectors. If, instead of forming a p^+ junction in n-type material, or an n^+ junction in p-type material, a wafer of n-type silicon or germanium is exposed to air for a few days an oxide layer is formed. Oxygen is electronegative and therefore an acceptor and so forms a p layer on the n-type material. A very thin layer of gold is evaporated on to the oxide surface and an aluminum contact is evaporated on to the back of the wafer. Prior to formation of the p layer, the edges of the wafer are etched in order to reduce surface leakage currents. The gold not only serves as a contact but, for reasons that are not altogether clear, facilitates the formation of a p-n junction. This type of detector is known as a *surface-barrier detector* although it is fundamentally no different from the rectifying-diode, or p-n, junctions already described. They are, however, characterized by very thin entrance windows of less than 10 nm of gold, depletion depths of 0.002 to 5 mm, large surface areas of up to 4 to 5 cm², and excellent energy resolution. In some cases accelerated boron ions have been implanted to establish the front contact. By comparison p-type silicon diffused p-n junction detectors are characterized by entrance windows (the thin n^+-type region) of 0.2 to 2 μm, depletion depths of 0.05 to 0.5 mm, and good resolution.

The various p-n junction detectors are widely used for the detection and energy spectrometry of charged particles such as low-energy electrons, protons, alpha particles and fission fragments. Their suitability for use as detectors of more penetrating radiations such as energetic electrons (such as beta particles) and photons is severely limited by their shallow radiation-sensitive charge-depletion regions. The need for detectors of these more penetrating radiations led to the development of large-volume detectors fabricated from high-purity and intrinsic material, to be described in the next section.

Lithium-drifted or p-i-n Detectors. Until recently, it has not been possible to produce semiconductor material of sufficient purity for the fabrication of high-purity high-resistivity detectors. In the case of germanium this would demand an impurity level of the order of 10^{10} electrically active atoms per cubic centimeter, or less than one such impurity atom per 10^{12} germanium atoms. Since the early 1970's it has been possible to produce such single crystals of germanium, but in the case of silicon considerably greater purification problems have been encountered, related mainly to its higher melting point. In 1960, however, E. M. Pell showed that lithium ions could be drifted interstitially through large volumes of silicon and germanium. Because lithium atoms act as donors, lithium can therefore be drifted into high-purity p-type silicon or germanium to compensate for residual acceptor atoms. This is accomplished by evaporating a lithium layer on to one face of a silicon or germanium crystal maintained at a temperature of around 400°C in a vacuum evaporator. The lithium is allowed to diffuse into the crystal for a few minutes in which time, depending on its

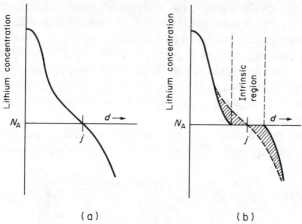

(a)   (b)

FIG. 6-21   Lithium concentration in p-type crystal as a function of depth, d, (a) before, and (b) after drifting.

duration, it will reach a depth of from 0.3 to 0.7 mm. This is illustrated in Fig. 6-21a where the concentration of lithium ions, N_D, is shown as a function of depth, d. The concentration of acceptor atoms, N_A, is also shown, and, at a depth j where $N_D = N_A$, a p-n junction is formed.

Having formed a p-n junction by diffusion, the next step is to drift the Li^+ ions into the crystal under the influence of a reverse-biased electric field (i.e. the positive voltage to the n^+ surface and negative to the p-type material). This is carried out with a reverse bias of from 500 to 1000 V at temperatures ranging from 0° to 100°C and for a period of from several days to a month, all these parameters depending on the size and material of the crystal.

As the Li^+ ions drift across the p-n junction they form complexes with the residual acceptor atoms which can therefore no longer accept electrons from the valence band, and the resistivity of the region increases as the formation of positive charge carriers (holes) is inhibited. Li^+ ions that have not been complexed continue to drift interstitially through the semiconductor lattice under the influence of the electric field, and a depletion region of very exactly compensated acceptor atoms is formed, as illustrated in Fig. 6-21b. Actually the process of compensation is self-adjusting in that if $N_A \gtrless N_D$ anywhere in the depletion region, the resulting space charge will produce an electric field that will tend to move Li^+ ions out of regions where they over-compensate acceptor atoms, and into regions where the acceptor atoms are under-compensated.

Thus the absolute value of the difference between the concentrations of the acceptor atoms and Li^+ ions, $|N_A - N_D|$, in the depletion region approaches zero and the resistivity of the semiconductor approaches that of intrinsic material.

The operation of such a detector, that is often called a p-i-n detector, is illustrated in Fig. 6-22. The operation of a high-purity germanium

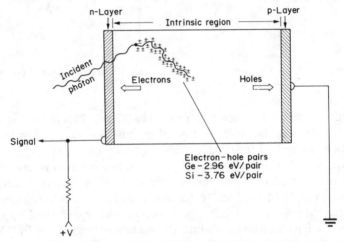

FIG. 6-22   Illustration of a p-i-n planar semiconductor detector, showing a typical photon interaction with the detector, and the movement of electrons and holes.

detector is exactly similar, but electrical contacts are usually formed by diffusing an n^+ layer of lithium on to one surface and forming a p-type contact, or a surface-barrier contact, on to the other. Such a detector in which the germanium is still very slightly p-type might now be designated as pπn.

For more explicit details of the preparation of the crystals and of the diffusion and drifting processes the reader is referred to the original paper by E. M. Pell and to a recent paper by R. C. Trammell both of which are cited in the bibliography of this chapter. The latter author also describes the final, so-called, "clean-up drift" at lower temperature and current, that is required to re-distribute Li^+ ions after the normal drift. This is necessary because, in the normal drift, the charge densities in the crystal, due to the higher currents, are also compensated by Li^+ ions.

Li^+ ions can be drifted to depths of the order of 5 mm in silicon and 2 cm in germanium. Because of the relatively high diffusivity of Li^+ ions in germanium, Ge(Li) detectors must be maintained and operated at low temperatures, normally that of boiling liquid nitrogen (77 K), to prevent the precipitation of lithium out of the crystal and loss of the depletion layer. As the diffusivity of lithium in silicon is fairly low, Si(Li) detectors can be operated at room temperature, although, normally, such detectors are operated at less than 100 K in order to reduce the leakage current which may be significant at higher temperatures for larger-volume detectors.

Planar and Coaxial Detectors. The detector such as that illustrated in Fig. 6-22, in which the lithium on a flat surface of the semiconductor crystal was drifted across the crystal, is usually called a *planar* detector. Larger sensitive, depleted, volumes can, however, be achieved by drifting radially in a cylindrical geometry as illustrated in Fig. 6.23. Such a detector is known as *coaxial*. Coaxial Ge(Li) detectors can be fabricated

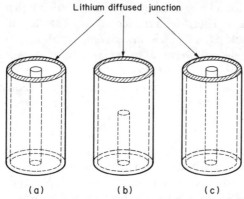

Lithium diffused junction

(a)   (b)   (c)

FIG. 6-23   Examples of three different types of coaxial (Ge(Li)) detectors. According to the method of fabrication, these are (a) open-ended coaxial; (b) closed-end coaxial; (c) hollow-core coaxial.

with volumes of up to $130 \, cm^3$ and can be solid or closed-end with p-type core (Fig. 6-23a and b) or hollow core (Fig. 6-23c). The solid-core detectors are fabricated by drifting lithium from the outside; the hollow-core by machining a germanium crystal and evaporating lithium on to the outside surface of the cylinder, or, less commonly, on to the interior surface of the hole and drifting inwards or outwards, respectively.

Energy per Electron-Hole Pair. The interaction of a charged particle or photon in a semiconducting crystal produces one or more electrons having energies considerably in excess of thermal energies. Such electrons rapidly (in the order of $10^{-12} \, s$) dissipate their energies in collisions within the crystal, creating more electron-hole pairs and phonons (quantized lattice vibrations). The average energy expended in creating one electron-hole pair, including that expended in competing processes, amounts to 2.96 eV in the case of germanium and 3.76 eV for silicon at 77 K. This is considerably less than the average energies of between 25 and 35 eV required to produce one ion pair in a gas-ionization detector, and of approximately 1000 eV to produce one photoelectron from the photocathode of the phototube of a scintillation detector. The resolution of a detector, that is, its ability to separate events that have deposited not too different amounts of energy in it, is related to the statistical fluctuations in its response. Such fluctuations are relatively much smaller for a large number than for a small number of detectable secondary events. Therefore a semiconductor detector producing some 10^5 electron-hole pairs for an ionizing event depositing 3×10^5 eV in its sensitive volume will have a better resolution than, say, a pulse ionization chamber that, for the same energy deposited, would produce only some 10^4 ion pairs (see section on Energy Resolution in Chapter 8.)

Calorimetry

Calorimetry finds little or no application in the field of nuclear medicine to which this text is primarily directed. Its principal use at the present time is in so-called nuclear-safeguards measurements, namely the control by accurate assay of nuclear-energy materials such as the isotopes of plutonium and of americium-241.

A calorimeter, a device for measuring quantities of heat or rates of energy emission, is used in many fields of chemistry and physics. Its use in radioactivity measurements is based on the fact that radiation energy which is absorbed in matter is ultimately degraded into heat energy. The calorimeter's greatest use is in measurements of the activities of α- and β-particle emitters, because it is much easier to absorb all the energy from these radiations (except for some of the bremsstrahlung) than that from γ-ray emitters. Essentially a calorimeter consists of a chamber that is thermally isolated from its surroundings. A source of heat is placed inside the chamber, and the energy output, or rate of energy emission, is determined by one of several methods. In the *adiabatic* type, as the temperature of the chamber rises due to absorption of energy, the temperature of an outer jacket is raised by an external source of heat so that there is no net transfer of heat between the inner chamber and the outer jacket. As this condition of equilibrium is maintained, the total energy input to the inner chamber, E, is related to the heat capacity, C, of the chamber and its contents, and to the change in temperature of the outer jacket, T, by the equation

$$E = C \, \Delta T.$$

In the *isothermal* type of calorimeter, the inner chamber is allowed to come to temperature equilibrium with its surroundings by removing heat at a uniform rate, such as by circulating water at a constant temperature in an outer jacket.

Another method suggested in 1910 by H. L. Callendar employs the *Peltier cooling* effect to remove the heat generated in the calorimeter.

The *twin-differential* type of calorimeter consists of two calorimeter chambers, sometimes called "thermels," which are constructed as nearly identical to each other as possible. The unknown source of heat is placed in one of the chambers. The other contains a coil, the current through which (i.e. the rate of heat loss) is adjusted so as to maintain zero difference in temperature between the two chambers. When this "null" condition is established, the rate of energy output of the unknown source of heat is equal to that from the electric heating coil.

Alternatively, in the last method, the two identical chambers, or thermels, may have identical heating coils wound around them, and their sensitivities in terms of the differential output from two resistance thermometers or thermocouples, also incorporated into the thermels, can be determined by means of a sensitive resistance bridge or current-measuring system. Calorimeters of this type have been brought to such a high degree of perfection by K. C. Jordan that he has been able to observe the decay of a 3-kCi source of tritium over a period of only three to four hours. Large-capacity calorimeters of this type are also used to assay the bulk waste remaining from the fabrication of nuclear-fuel elements.

The National Bureau of Standards Bunsen ice calorimeter, in which the rate of energy dissipation is balanced against the rate that ice is melted, has been used to measure the power output from plutonium sources of the order of 1 watt. A National Bureau of Standards microcalorimeter has also been used to standardize nickel-63 and, in conjunction with mass spectrometry (using the relation $\lambda = (1/N)\,dN/dt$), to determine its half life.

Calorimetric methods have an inherent advantage of relatively simple and precise instrumentation, compared with the complex electronic equipment required for most of the other quantitative detectors. It is, of course, necessary to know the energies of the α particles and the mean energies of the β particles in order to relate the rates of energy emission to disintegration rates. The fundamental relationship is that 1 MeV is equal to 1.602×10^{-6} erg, from which one can derive that Q, in watts, is equal to $1.602 \times 10^{-13} \overline{N E}$, where $\overline{N}$ is the mean number of transitions occurring per second, and $\overline{E}$ is the average energy emitted, in MeV, per transition.

Calorimeters are now widely used to assay for the amounts of special nuclear materials, such as the isotopes of plutonium, by means of a measurement of the rate of energy emission combined with a knowledge of the isotopic abundances and the *specific power* (i.e. the rate of energy emission per unit mass) of each isotope in a sample.

References

Adams, F. and Dams, R. (1970) *Applied Gamma-Ray Spectrometry*, 2nd ed. (Pergamon Press, Oxford and New York).
Birks, J. B. (1964) *The Theory and Practice of Scintillation Counting* (Pergamon Press, Oxford).

Bransome, E. D., (Ed.) (1970) *The Current Status of Liquid Scintillation Counting* (Grune & Stratton, New York).

Campion, P. J. (1973) Spurious pulses in proportional counters: a review, *Nucl. Instrum. Methods, 112*, 75.

Corson, D. R. and Wilson, R. R. (1948) Particle and quantum counters, *Rev. Sci. Instrum., 19*, 207.

Geiger, H. and Müller, W. (1928) Das Elektronenzählrohr, *Phys. Z., 29*, 839.

Geiger, H. and Müller, W. (1929) Technische Bemerkungen zum Electronenzählrohr, *Phys. Z., 30*, 489.

Goulding, F. S. and Pehl, R. H. (1974) Semiconductor radiation detectors, Chap. III.A in *Nuclear Spectroscopy and Reactions*, Part A, J. Cerney (Ed.) (Academic Press, Inc., New York).

Hibberd, R. C. (1968) *Solid-State Electronics* (McGraw-Hill, New York).

Hine, G. J., (Ed.) (1967) *Instrumentation in Nuclear Medicine*, Vol. 1 (Academic Press, New York).

Hoffer, P. B., Beck, R. N., and Gottchalk, A. (Eds.), (1971) *Semiconductor Detectors in the Future of Nuclear Medicine* (Society of Nuclear Medicine, New York).

NCRP (1978) National Council on Radiation Protection and Measurements. *A Handbook of Radioactivity Measurements Procedures*, NCRP Report No. 58 (National Council on Radiation Protection and Measurements, Washington).

O'Kelley, G. D. (1962) Detection and measurement of nuclear radiation, NAS-NS 3105 (National Academy of Sciences, Washington).

Pell, E. M. (1960) Ion drift in a *n-p* junction, *J. Appl. Phys., 31*, 291.

Price, W. J. (1964) *Nuclear Radiation Detection*, 2nd ed. (McGraw-Hill, New York).

Snell, A. H. (Ed.), (1962) *Nuclear Instruments and Their Uses*, Vol. 1 (John Wiley & Sons, New York).

Staub, H. H. (1953) Detection Methods, Part I in *Experimental Nuclear Physics*, E. Segre, (Ed.) (John Wiley, New York; Chapman & Hall, London).

Steyn, J. J. and Nargolwalla, S. S. (1973) Detectors, Chap. 4 in *Instrumentation in Applied Nuclear Chemistry*, J. Krugers (Ed.) (Plenum Press, New York).

Taylor, J. M. (1963) *Semiconductor Particle Detectors* (Butterworths, Inc., Washington).

Thewlis, J. (Ed.-in-Chief) (1962) *Encyclopaedic Dictionary of Physics* (Macmillan, New York; Pergamon Press, Oxford and London).

Trammell, R. C. (1978) Semiconductor detector fabrication techniques, *IEEE Trans. Nucl. Sci., NS-25*, 910.

7 Electronic Instrumentation

THE output signals from the various radiation detectors described in Chapter 6, except for the Geiger–Müller counter, are generally quite small, typically in the range of 10^{-10} to 10^{-15} C per event. These signals, except for the direct-current mode of operation of an ionization chamber, are current pulses which flow for a very brief time, that is, pulses of charge (each charge pulse being the time integral of the current). The area of each pulse, that is the total charge, is, for most detectors, directly proportional to the energy deposited within the sensitive volume of the detector.

Prior to any analysis of a detector pulse, such as simple counting or pulse-amplitude measurements, the pulse is normally shaped and amplified to produce a signal that is compatible with the requirements of the analyzing instrument. Shaping of a pulse is effected during amplification by the use of appropriate gain-stage coupling networks.

A complete understanding and appreciation of the complexities and subtleties of most of the electronic instruments that are used in conjunction with radiation detectors is beyond the scope of this book. It may well be, however, that the reader has no prior knowledge of electronics whatsoever. We shall therefore discuss the various instruments of interest mainly on a functional basis, with but minimal reference to the electronic details. Readers interested in these details and other subtleties of the instruments discussed should consult appropriate references given in the bibliography of this chapter.

Pulse Shaping

The output pulse from a typical detector is a current or voltage pulse having a very short *rise time* and a comparatively long time of fall. The rise time, namely that required to go from 10 to 90 per cent of the peak amplitude of the pulse, is determined by, and is nearly equal to, the duration of the flow of current within a detector, or the phototube of a scintillation detector. The duration of this current flow is determined by the type of detector and the nature and energy of the ionizing-radiation event that has been detected, and can vary from as short a time as 0.1 ns for a plastic scintillator to times in excess of 5 μs for certain gas-filled ionization detectors. The corresponding decay time, namely the time required for the pulse to decay exponentially to 1/e or 37 per cent of its peak amplitude, is determined by the time constant of the detector circuit (i.e. the product of the resistance and capacitance between the terminals of the detector). Such pulses with rise times less than about 1 μs and decay times greater than 50 to 100 μs are commonly referred to as *tail pulses*.

One of the essential requirements of detector pulse measurements is that the individual pulses from the detector must be analyzed independently of any interference from preceding pulses. With the long detector time constants, typically from 50 to 100 μs characteristic of most detector systems, the time required for a pulse to decay to the baseline (zero-volt) level is often of the order of 1 ms or longer. At counting rates in excess of a few hundred events per second, *pile-up* of the pulses from the detector will begin to occur. This phenomenon of pulse pile-up, caused when a pulse overlaps with the tail of a preceding pulse, is illustrated in Fig. 7-1a. When a pulse piles up on the tail of a preceding pulse, or pulses, it is distorted with a consequent change in its own amplitude. At higher rates the pile-up of the individual pulses can exceed the dynamic input range of one or more of the succeeding amplifier stages, resulting in their paralysis.

The adverse effects of pulse pile-up can be greatly reduced if the slowly-decaying pulses from the detector are shortened or *clipped*, as illustrated in Fig. 7-1b, in order to terminate them in a time that is short compared with the average time interval between the pulses. This can best be achieved by differentiating the pulse in order to produce a shaped pulse that is proportional to the derivative of the input pulse (i.e. the output pulse from the detector). One of the simplest and most commonly used circuits for performing this function is an *RC differentiating circuit*, such as that shown in Fig. 7-2a, together with its response to a step input pulse. The voltage relationship between the input and output signals, as

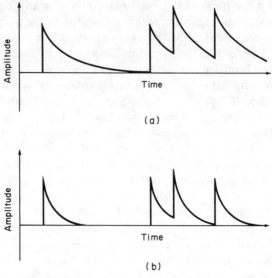

FIG. 7-1   Pulses from a typical radiation detector: (a) at the amplifier input; (b) after differentiation with a short time constant. Note: It is assumed the undershoot normally associated with the differentiated pulses has been removed (see section on amplifiers).

a function of time, is also given. Thus, as shown in Fig. 7-2a, pulses are clipped and decay with a time constant, τ, that is determined by the product of the resistance and capacitance of the circuit.

This circuit also functions as a *high-pass filter*, namely one that attenuates any low-frequency-noise components that may be super-imposed on the desired signal. The differentiation circuit thus performs two important signal-processing functions, namely (i) by clipping the pulse, it reduces the probability of pulse-amplitude, or "pulse-height", distortion due to pile-up, and (ii) it improves the signal-to-noise ratio to the extent that the low-frequency noise component of the signal is attenuated by the circuit.

The output pulse produced by a single differentiating circuit is not, however, especially well suited for subsequent signal-processing, for several reasons, One is that the pulse shape, particularly the narrow width at its peak, is not well suited for such subsequent processing by

pulse-height-analysis instruments that require, for optimum performance a more gradual change in slope near the peak. Another is that the
signal-to-noise ratio of the pulse could be further improved by removing,
or attenuating, the high-frequency-noise components.

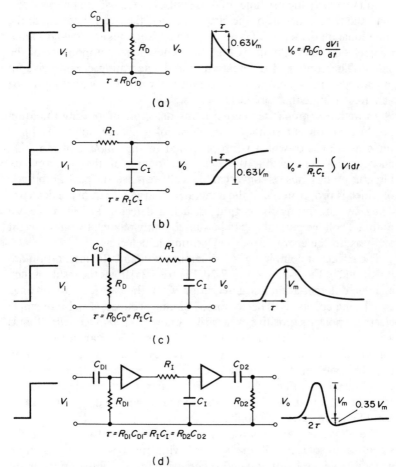

FIG. 7-2   *RC* Shaping networks: (a) single *RC* differentiation; (b) single *RC* integration;
(c) combined differentiation-integration; (d) combined double differentiation and single
integration. V_m is the maximum amplitude of the output pulse.

Additional shaping of the pulse can, however, be achieved by the use of an *integrating circuit*, the output of which is proportional to the integral of the input signal. A typical *RC* integrating circuit is shown in Fig. 7-2b, together with the response of the circuit and the corresponding relationship between the input and output signals. The integrating circuit increases the rise times of those pulses that have rise times shorter than the time constant of the circuit, so that their rise times equal the time constant of the circuit. This circuit also functions as a *low-pass filter*, namely one that attenuates high-frequency-noise components of the signal. The combined differentiating and integrating circuits give an improved signal-to-noise ratio compared with that which could be achieved using either circuit alone.

As we have emphasized earlier many detectors of ionizing radiation produce an output signal, in the form of a charge pulse, the time-integrated area of which is proportional to the amount of radiation energy deposited within the sensitive volume of the detector. By integration this pulse is converted to a voltage pulse, the peak amplitude of which is proportional to the integrated charge output of the detector. Thus, by integrating the output of such a detector, at least once, we obtain a voltage pulse that can be subsequently related, by pulse-height analysis, to the energy deposited within the detector.

The effects of combining *RC* differentiation and *RC* integration are illustrated in Fig. 7-2c and Fig. 7-2d. Figure 7-2c shows the result of such a single differentiation and integration when the two *RC* circuits have equal time constants. The combined circuit produces a pulse of single electrical polarity (positive or negative) known as a *unipolar pulse*, that is asymmetric about the peak with respect to time. The triangles, shown in this circuit and in circuit of Fig. 7-2d, represent the electrical isolation required between the shaping circuits, or networks as they are often called, in order to achieve the desired circuit functions. The isolation represented by these triangles is normally achieved by the use of amplifier-gain stages, or "buffers."

The *RC* time constants used are normally selected to optimize the signal-to-noise ratio of the system, with the best noise reduction generally achieved using equal differentiating- and integrating-circuit time constants. The signal-to-noise ratio of a singly differentiated (unipolar) pulse improves with the addition of further integrating networks, an optimum limiting result being a Gaussian-shaped pulse that would be produced by an infinite number of such networks.

If the step input is doubly differentiated by two differentiating circuits and integrated by a single RC integrating circuit, as shown in Fig. 7-2d, a pulse having both positive and negative values and a single baseline crossing (*zero crossover*) is produced. Such a pulse is known as a *bipolar pulse*. The areas of the positive and negative lobes of the bipolar pulse are equal to each other, their relative amplitudes, however, being determined by the time constants of the RC shaping circuits. In the example given in Fig. 7-2d for circuits having equal time constants, the amplitude of the negative lobe of the pulse is 0.35 times that of the positive lobe.

For any input waveform, other than the step function used in the examples of Fig. 7-2, an RC differentiating network will not produce a unipolar pulse. Thus, differentiation of typical tail pulse, such as that shown in Fig. 7-1a, will produce a pulse with a positive lobe, the amplitude of which will be proportional to the amplitude of the input pulse, followed by a negative lobe or *undershoot*, the area of which will equal that of the positive lobe. The amplitude and decay time of the undershoot of the bipolar pulse are determined by the longer of the two time constants. Similarly, if a tail pulse is doubly differentiated an overshoot follows the negative lobe of the bipolar pulse produced by the second differentiation. In practice, such overshoots and undershoots should be avoided because of the increased probability of pile-up and consequent amplitude distortion of succeeding pulses. Several methods have been developed to eliminate or minimize these potential sources of pulse-amplitude distortion, and these will be discussed in the section on amplifiers.

So far we have considered the shaping of pulses (differentiation and integration) solely in terms of RC shaping circuits. Other circuits, such as *resistance-inductance* (L/R) networks, are also used in a similar manner to perform the desired differentiating and integrating shaping functions. Such circuits will give shaping that is comparable with that obtained using RC networks, the time constant of an L/R circuit being the ratio of the inductance L, in henries, to the resistance R, in ohms.

Another popular method for obtaining a similar result to differentiation is to use a *delay-line* circuit. A delay line is simply a circuit component that delays the propagation of a signal between the input and output terminals of the component for a time t_d, that ranges from a few nanoseconds to hundreds of microseconds. Such delay lines can be used in several ways to convert an input signal into an output pulse of controlled duration. One way is to feed the signal into a delay line that is short-circuited at one end, matching the impedance of the signal source with that of the delay line, in the manner shown in Fig. 7-3a for a step-function input signal. After a time equal to twice the delay time, the pulse

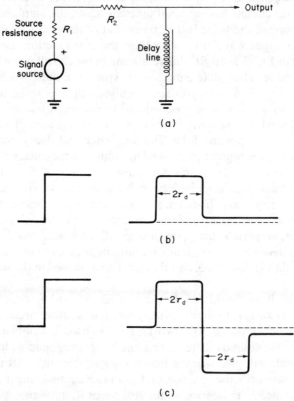

FIG. 7-3   Delay-line differentiation: (a) single-delay-line "differentiation" circuit with R_2 chosen such that $R_1 + R_2$ is equal to the impedance of the delay line; (b) single-delay-line "differentiated" signal showing "rear porch" due to delay-line attenuation; (c) comparable double-delay-line "differentiated" signal (bipolar pulse).

that is reflected back from the short-circuited end of the delay line with inverted voltage interacts with the input signal and produces a unipolar square pulse of width $2t_d$. This is illustrated in Fig. 7-3b for an ideal delay line. In actual practice, a delay line attenuates the delayed signal so that the cancellation of the input signal is incomplete. This results in what is called a *rear porch* which is comparable with the undershoot or overshoot of an RC-shaped pulse.

Another technique of delay-line "differentiation" is to use a delay line that is terminated by its characteristic impedance at both ends, so as to delay the input signal without reflection. The delayed signal is then subtracted from the input signal in a *difference amplifier*. This technique allows for the possibility of slightly attenuating the input signal at the difference-amplifier input to compensate exactly for attenuation of the delayed signal in the delay line, thereby eliminating the rear porch from the shaped output pulse.

Input signals may also be doubly "differentiated" by means of delay-line circuits in a manner analogous to that of RC double differentiation. Delay-line double "differentiation" of a step-function input signal will produce a bipolar output pulse such as that illustrated in Fig. 7-3c.

Preamplifiers

In most radiation measuring systems, operated in the pulse mode, a preamplifier is incorporated between the detector and the main amplifier. Its purpose is to provide one or more of the following functions: (i) first-stage amplification of the detector output pulse; (ii) preliminary pulse shaping; (iii) matching of the detector output impedance with that of the input signal cable to the main amplifier; and (iv) charge-to-voltage conversion of the detector output pulse. For reasons that will be explained later, the preamplifier, for best performance, must be located within a short distance, say 30 cm or less, from the detector.

A block diagram of a typical ac-coupled *charge-sensitive preamplifier* is shown in Fig. 7-4. Each of the various stages of this type of preamplifier and their respective functions are also indicated in this figure. We have chosen this type of preamplifier as an example because of its

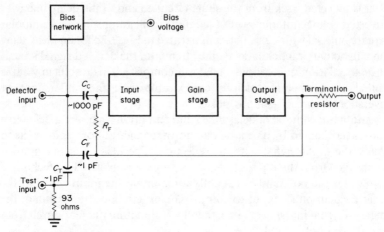

FIG. 7-4   Block diagram of a typical low-noise charge-sensitive preamplifier (after Ayers, 1973).

compatibility, and consequent widespread use, with most detector systems.

In order to understand the operation of this preamplifier, let us consider each function of the preamplifier individually. Of these, the amplification function can itself be subdivided into the three stages shown in Fig. 7-4, of which the input stage and the output stage are of particular significance in determining the overall performance of the preamplifier.

We will first consider the amplification function of the preamplifier as a whole, as determined by the combined operation of the three stages (input, gain and output), together with the "feedback" capacitor C_f and the resistor R_f, shown in Fig. 7-4. This can best be done by referring to the electrical equivalent of this functional block diagram, shown in Fig. 7-5. In this figure the three separate amplifier stages are represented by a triangle, with *negative feedback* provided by the parallel combination of the "passive" linear circuit elements R_f and C_f. This simple and convenient representation of an amplifier conceals the often complex circuit of the amplifier itself; those readers interested in the operation of an amplifier at the discrete-component level should consult the references cited at the end of this chapter. It will also be assumed that the amplifier is "ideal" in the sense that there are no time delays in its solid-state transistor components. The terms "closed loop" and "open loop" will be used to describe the operation of the circuit with and without the feedback loop, respectively. The designation "-A" on the amplifier indicates the inversion of the output signal with respect to the input signal, and the positive and negative signs to the left of the amplifier "triangle" signify the noninverting and inverting input terminals. In the circuit depicted in Fig. 7-5, the noninverting terminal is connected through the amplifier case to ground potential.

With negative feedback, the amplifier acts as a sensor and operates to reduce the input signal, across its input terminals, to zero. To achieve this, a fraction, f, of the inverted amplifier output is fed back and subtracted from the input signal. The size of f is determined by the feedback element. Let us consider an input charge Q_i from the detector. This charge will give rise to an input voltage, V_i, to the amplifier, equal to Q_i/C_i, where C_i is the open-loop input capacitance and is equal to the sum of the detector and stray capacitances, $C_d + C_s$.

With feedback, or closed-loop operation, the output signal, V_0, will be equal to $-V_i(1 - f)A$ where $-A$ is the open-loop amplification. The

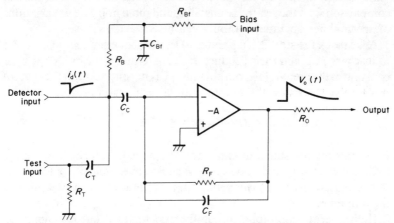

FIG. 7-5   Equivalent circuit of charge-sensitive preamplifier shown in Fig. 7-4. Always present, but not shown, are the detector capacitance and resistance and stray input capacitance and resistance.

consequent voltage across C_f will now be equal to $V_0 - V_i$, or $V_0(1 + 1/(1 - f)A)$ if Q_i. and hence V_i, is negative.

As, however, the feedback element operates to reduce Q_i to zero, it is necessary that

$$C_f V_0(1 + 1/(1 - f)A) = Q_i. \qquad (7\text{-}1)$$

Because A and f are typically of the order of 5000 and 0.1, respectively, $1/(1 - f)A$ will be of the order of 0.001, and can be neglected in Eq. 7-1, in which event

$$V_0 = Q_i/C_f. \qquad (7\text{-}2)$$

In actual practice, however, no amplifier is ideal, so that small corrections, incorporating the open-loop amplification and the feedback ratio, f, must be made to this result. Nevertheless, the voltage gain of a charge-sensitive preamplifier with negative feedback is essentially independent of the gain of the amplifier stage itself, but is, for a given detector circuit, dependent only on the passive linear circuit element C_f that is used to provide the negative feedback. This feedback element is quite stable with respect to time and temperature variations in comparison with the relatively unstable, and often non-linear, variations of an amplifier operated without negative feedback.

A feedback resistor, R_f, in parallel to the feedback capacitor C_f, serves to discharge C_f, and their product, $R_f C_f$, determines the discharge time constant of the preamplifier output pulse. Taking into account the effect of this feedback resistor, the output pulse as a function of time, t, is

$$V_0(t) = -Q_i/C_f e^{-t/R_f C_f}. \qquad (7\text{-}3)$$

From this relationship, one can readily appreciate that the value of the feedback capacitor should be as small as possible, typically of the order of 1 picofarad, in order to maximize the amplitude of the output pulse of the amplifier.

If the decay time constant of the preamplifier output signal, as determined by the product $R_f C_f$, is much greater than the detector charge collection time T_c, then the *peak* value of the preamplifier output voltage, $V_0(t)$ as given by Eq. 7-3, can be determined from

$$V_{0,p} = - (1/C_f) \int_0^{T_c} i_d(t)\, dt$$

$$= - Q_i/C_f, \tag{7-4}$$

where $i_d(t)$ is the detector current as a function of time (as shown in Fig. 7-6a), Q_i is the corresponding total charge produced in the detector, and T_c is the charge-collection time (i.e. the total time for which the current $i_d(t)$ flows). As is to be expected, the peak output voltage given by Eq. 7-4 is equal to that given by Eq. 7-2, where no discharge of the integrating feedback capacitor occurs. This proportionality between the integrated charge from the detector and the amplitude of amplifier output pulse, given by Eq. 7-4, is based on the assumption that the decay time of the output pulse is much greater than the detector charge-collection time, is of considerable significance with regard to subsequent signal processing. The amplitude of the resultant voltage pulse, that can be measured by rather straightforward methods, can be related directly to the energy deposited in the detector. The forms of the current and voltage input signals and of the voltage output signals as functions of time are illustrated in Fig. 7-6. The time relationship between detector charge collection and the growth of the preamplifier output voltage pulse is illustrated in Figs. 7-6a and 7-6b. The corresponding decay of the preamplifier output pulse is indicated in Fig. 7-6c, with an appropriately lengthened time scale.

For detailed derivations of the relationships between the preamplifier input and output, and for descriptions of the associated electronic circuitry, the references cited at the end of this chapter should be consulted, particularly H. H. Chiang (1969) and J. B. Ayers (1973). These references also treat, in considerable detail, many of the topics, such as voltage-sensitive preamplifiers, dc-coupling to the detector, and count-rate-related effects, that have been omitted here but which might be of interest to the reader.

Having described the operation of the amplification function of the preamplifier as a whole, we will now consider the operation of the individual input and output stages, shown in Fig. 7-4, in terms of their importance in determining the overall performance of the preamplifier.

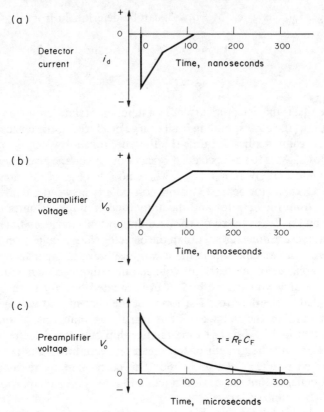

FIG. 7-6 Relationships between the input and output signals of a charge-sensitive preamplifier as a function of time.

Detectors of ionizing radiation are characterized by high output impedances. In order to amplify signals from such high-impedance sources, without loading the signal source and distorting the signal, an amplifier is required that has an input-stage impedance that is greater than that of the signal source. The use of a field-effect transistor (FET) in the amplifier-input stage provides such a high-impedance input, while, at the same time, providing the lowest possible noise contribution to the signal, when compared with such alternative devices as vacuum tubes and bipolar transistors. Presently available FET's are, for example, quieter than the best available vacuum tubes by a factor of at least 5 to 10 at room temperatures. The excellent noise performance of an FET input stage can be further improved by cooling the FET and its associated circuit components.

For many detector-system applications, such as the processing of large-amplitude low-resolution signals from scintillation detectors or for the simple counting of events, we do not require the high-quality noise performance of FET input-stage preamplifiers. For such applications, a low-cost voltage-sensitive preamplifier with a relatively noisy bipolar transistor-input stage provides more economical but adequate performance.

The output signals from a preamplifier are often transmitted to a main amplifier over distances of 30 m or more, through low-impedance shielded coaxial cables. Such cables (e.g., RG-62/U and RG-58/U cables with nominal impedances of, respectively, 93 ohms and 52 ohms) are used because of their reduced tendency to pick up unwanted signals (noise), as compared with high-impedance transmission lines. However, the use of low-impedance signal cables requires a preamplifier with a low-impedance output stage that is capable of high-current "drive" in order to prevent attenuation of the signal. An output stage used to increase the power or current-handling capability of a circuit is often referred to as a *driver* or *output driver stage*. In present-day transistorized preamplifiers, a bipolar transistor in an *emitter-follower* configuration (described in many of this chapter's references) is used as the output driver stage to provide the required transformation from high to low impedance and high-current-drive capability. This emitter-follower circuit is designed so that the output impedance of this stage matches the characteristic impedance of the output-signal cable as closely as possible.

The need to use a preamplifier with the capability of supplying sufficient current to transmit a pulse through a considerable length of low-impedance cable, without significant attenuation of the signal can best be illustrated by means of a specific example: Consider a detector, having a capacitance of 10 to 25 pF, that is located, without any preamplifier, 30 m from the first amplifier stage and connected to it by a coaxial cable with a capacitance of $75 \, pF \, m^{-1}$, for a total cable capacitance of 2250 pF. As the magnitude of the input voltage pulse at the first amplifier stage, for a given charge impulse Q_i from a detector, is given by Q_i/C_i where C_i now includes the capacitance of the signal cable, it is apparent that the amplitude of the voltage pulse produced by a given charge input is greatly reduced by this added input capacitance. In our example, the voltage pulse produced would be reduced by a factor of about one hundred with that which could be obtained by connecting the detector directly to a preamplifier without the cable.

In addition to the amplifier circuit itself, several other circuit functions are normally included as part of a typical preamplifier. These are the detector bias network and test input, shown in Figs. 7-4 and 7-5.

The detector bias network is simply a two-component circuit, the function of which is to provide the bias voltage necessary for the proper operation of the detector and to block this voltage from being "seen" by the input stage of the preamplifier. The decoupling capacitor C_c is used to prevent the sensitive low-voltage input of the preamplifier from being destroyed by the relatively high voltages used to bias most detectors. The second component, a bias resistor R_b, limits the current supplied to the detector by the external voltage supply. The value of this resistor, typically between 10^6 and 10^{10} ohms, depending on the magnitude of the sum of detector leakage and signal currents, should be as large as possible in order to minimize noise contributions to the detector from the voltage supply and the bias network itself. The value must not, however, be so large as to produce a significant voltage drop across R_b when supplying the necessary operating current for the detector under normal conditions.

For most commercially available preamplifiers, the built-in detector bias network has an upper-voltage limit of about 5000 V.

The preamplifier test input is provided for the purpose of introducing an external voltage tail pulse, either to test or to monitor the operation of the preamplifier or the entire measuring system, excluding the detector. The charge input to the preamplifier, Q_t, for a given voltage input, V_t, is $C_t V_t$, where C_t is the capacitance the test input capacitor.

In conclusion, the features and operation of a typical charge-sensitive preamplifier have been discussed in some detail. This type of preamplifier may be satisfactorily used with almost all types of detectors. Where the capacitance of the detector varies with temperature, as is the case for most semiconductor detectors operated at room temperature, the performance of a charge-sensitive preamplifier is superior to all other types. This is due to the fact that the amplitude of the output signal from this type of preamplifier is almost completely independent of the input capacitance, and is a linear function of the input charge.

There are other types, such as current-sensitive and voltage-sensitive preamplifiers. The former are not normally useful because of their characteristically low input impedances which are not compatible with high-impedance detector outputs. The input impedance of a voltage-sensitive preamplifier can, however, be matched to that of a detector, and this type may therefore be successfully used with many detectors.

The voltage-sensitive preamplifier is quite similar to the charge-sensitive type. The main difference is that the feedback element in a voltage-sensitive amplifier is a resistor rather than a capacitor. Because the voltage-sensitive amplifier is extensively used in the main amplifier, its features and operation will be discussed in the following amplifier section.

For detectors such as cooled semiconductor detectors or proportional counters, either the voltage-sensitive or charge-sensitive type of preamplifier may be successfully used with comparable performance. On the other hand, for those detectors with large signal outputs, such as various scintillation detectors (with phototubes) and Geiger–Müller counters, the signal may be too large for typical charge-sensitive preamplifiers. In such cases the voltage-sensitive preamplifier with unit or very low gain may be the ideal type, its main use being to match the output impedance of the detector to the impedance of the signal cable.

Amplifiers

The voltage-sensitive nuclear pulse amplifier, or linear amplifier as it is often called, completes the amplification of the detector signals from the preamplifier to amplitudes appropriate for subsequent pulse analysis. To do this the amplifier is designed to provide continuously adjustable linear gain over a wide range. The overall gain of most nuclear pulse amplifiers ranges from about five or ten to one or two thousand.

The other function of the amplifier, and one that is critical in achieving optimum performance, is that of shaping the signal pulse. The typical tail pulse from a preamplifier (Fig. 7-6) is reshaped within the amplifier to

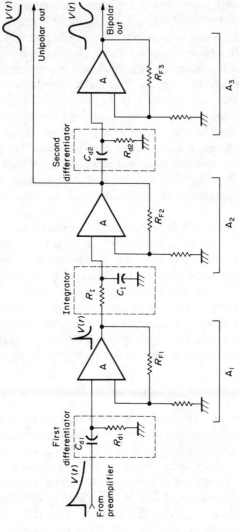

FIG. 7-7   Simplified diagram of a typical linear nuclear pulse amplifier with RC integration and both single and double differentiation of the output signal.

produce a much narrower pulse with slower rise time and a much faster fall time. The shape of the amplifier output pulse is dictated by considerations of signal-to-noise and compatibility with subsequent pulse-analysis electronics as determined by the purpose of the measurement. These two pulse-shaping criteria usually conflict with each other; thus the shaped pulse produced by the amplifier is usually a compromise between the shape for optimum signal-to-noise ratio and that best adapted to the pulse-analysis electronics for a given set of measurement requirements and conditions.

Many of the concepts necessary for describing the operations and the functions of a linear amplifier have already been described in the preceding sections on pulse shaping and preamplifiers. With that background, we can now describe the operation of a typical linear pulse amplifier such as that shown diagrammatically in Fig. 7-7. In this simplified diagram we see that a typical linear amplifier is composed of multiple gain stages (A_1 through A_3 in the figure) with various interstage coupling networks that function either as differentiators or integrators.

One of the primary requirements of a linear amplifier is that it should be stable; that is, its gain, once set, should be as constant as possible. The gain of a simple amplifier can be affected by changes in supply voltage, aging of components, and temperature variations. To minimize these unavoidable effects, the technique of negative feedback, which we have already discussed in the case of a charge-sensitive preamplifier, is also extensively used in the design of voltage-sensitive linear amplifiers. Consider one of the amplifier stages A_1, A_2 or A_3 shown in Fig. 7-7, with a negative input signal V_i, as shown in Fig. 7-8. Let us assume that in the absence of feedback, the inverting amplifier has a gain of $-A$, so that the output signal V_0 would be equal to $-AV_i$. With the introduction of

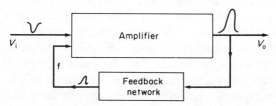

FIG. 7-8   Block diagram of an amplifier with negative feedback, f.

feedback, a fraction f of the output is fed back and subtracted from V_i, so that

$$V_0 = A(V_i - fV_0). \tag{7-5}$$

Solving for V_0, we obtain

$$V_0 = AV_i/(1 + fA). \tag{7-6}$$

As the gain of an amplifier is the ratio of the amplitudes of the output to input signals, the closed-loop gain is now given by

$$V_0/V_i = A/(1 + fA). \tag{7-7}$$

As, in general, Af $\gg$ 1, the closed-loop gain is seen to be, to a very close approximation, equal to 1/f. Inasmuch as the value of f, the feedback factor, of a voltage-sensitive amplifier is determined by the value of a resistance that can be made to remain quite invariant with temperature and time, so will the closed-loop gain remain invariant.

It can be shown that negative feedback also aids in preserving the *linear* characteristic of the amplifier. This is an essential requirement when spectral analyses are to be performed with, for example, proportional counters or semiconductor detectors. Just as the outputs of these detectors are directly proportional to the amounts of radiation energy deposited in them, so must the output signal from the amplifier be linearly related to the input signal, in order to take advantage of the linear characteristics of these radiation detectors.

The amplifier diagrams in Fig. 7-7 have been simplified by the omission of a number of circuit functions that are normally included in this type of linear pulse amplifier, such as variable gain controls, variable shaping time constants, and baseline restoration. Adjustable gain is typically provided by external variable control of the amount of negative feedback in one or more of the amplifier stages or, alternatively but with increased noise, by fixing the gains of the individual stages and varying the overall gain of the amplifier by means of externally variable input-signal attenuation. Similar external control (usually in the form of switch settings) is provided for the selection of suitable time constants for the interstage coupling networks that shape the output pulses.

As was stated in the section on shaping, the differentiation of a typical preamplifier output pulse (the tail pulse in Fig. 7-9a) within the linear

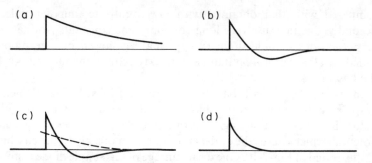

FIG. 7-9   Pole-zero compensation of differentiated pulse undershoot: (a) original tail pulse; (b) after passing through a coupling network of short time constant (differentiator); (d) resultant pole-zero compensated pulse (from Hatch, 1973).

amplifier will produce a pulse with an undesirable undershoot (Fig. 7-9b). This undershoot can, however, be removed by a technique known as *pole-zero cancellation* (see Fig. 7-9). In this technique a proper proportion of the original input pulse (Fig. 7-9c), as determined by a large variable resistance in parallel with the differentiator coupling capacitor, is added to the output pulse from the differentiating circuit. When properly adjusted, the resulting output pulse from this coupling network is a perfect tail pulse (Fig. 7-9d), with the short decay time constant characteristic of the differentiator and with no undershoot.

Amplifiers always have coupling capacitors and bypass capacitors that are charged and discharged by passing pulses. For unipolar pulses, these coupling networks produce an undershoot with a decay time determined by the RC time constant of the network. This undershoot, however, degrades the energy resolution of the system at high counting rates. For a high-resolution system capable of, say, 0.1 per cent resolution, coupling capacitors will always have the effect of degrading the resolution for count rates greater than $1000\,s^{-1}$.

Several methods may be used, however, to reduce the loss of energy resolution at high counting rates. First, bipolar shaping may be used because bipolar pulses (the two lobes of which have equal areas) leave a net zero charge when they pass through the coupling capacitor, and, therefore, do not produce long undershoots. On the other hand, there is a loss of about 40 per cent in the low-count-rate signal-to-noise ratio compared with that obtained with low-count-rate unipolar pulses. Secondly, one may use very long coupling time constants (so that the amplitude of the undershoot is very small), followed by a technique called *baseline restoration* that very quickly returns the undershoot to the baseline.

In its simplest form a baseline restorer is a biased diode. When a unipolar pulse returns to the baseline and goes into undershoot, the diode becomes conducting, shorting the signal nearly to ground potential which causes the undershoot to return quickly to the baseline, as illustrated in Fig. 7-10. One disadvantage of this simple diode restorer is that it distorts low-amplitude pulses, but this can be overcome by the use of a more sophisticated *amplified-diode-restorer* circuit called an *active restorer*.

A good amplifier should have low *equivalent* input noise, for which every amplifier has an associated *noise figure*. This is usually expressed as the amplitude of the noise pulse at the output divided by the gain. This noise arises from (i) thermal noise due to the random movements of

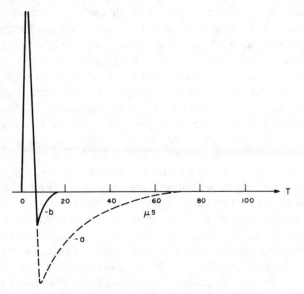

FIG. 7-10   Pulse shape (a) before and (b) after baseline restoration (from Hatch, 1973).

charged particles, (ii) shot noise due to the discrete nature of charges in a non-metallic medium, (iii) hum from the amplifier power supply, and (iv) pickup from nearby sources of electromagnetic radiation, such as radio and television transmitters or faulty fluorescent lamp fixtures. If an amplifier has a noise level of $100\,\mu V$, the lowest level of detectable signal amplitude from the detector will also be approximately $100\,\mu V$. Linear amplifiers with noise levels less than $15\,\mu V$ are now commercially available.

A well-designed amplifier should have good *overload* characteristics. Consider the situation when one is using a proportional counter to measure the activity of a β-emitting radionuclide. The output pulses from the detector will range in amplitude from the microvolt level to the order of a volt, in consequence of the continuous spectrum of β-particle energies. If one wishes to count all pulses of $15\,\mu V$ and larger, and the counting circuit only registers input pulses of, say, at least 0.015 V, then the amplifier should operate at a gain of $0.015/(15 \times 10^{-6})$, which is 10^3. Linear nuclear pulse amplifiers are capable of producing output pulses of 10 V. With the gain set at 10^3, any input pulse with an amplitude greater than 10 mV will *saturate* the amplifier. That is, no matter how much larger than 10 mV the input signal is, the amplitude of the amplifier output pulse will be just 10 V. If an input signal has an amplitude of 100 or 1000 mV, the amplifier is said to be driven to 10 or 100 times *overload*, and under these conditions the amplifier can be paralyzed by either of these pulses for periods of 20 to 50 μs or even longer. This has the effect of introducing into the circuit a variable *dead time*, which makes it impossible to obtain accurate activity measurements. Amplifiers can be designed, however, to minimize this paralysis, and even when any of the amplification stages, except the first, is driven into the saturation or overload condition, the resulting dead time can be limited to 5 or 10 μs. In the event of first stage saturation of the amplifier, it may be paralyzed for much longer times, because the long tails of the larger-amplitude input pulses from the preamplifier will keep the first stage in saturation until the pulse slowly decays below the saturation level.

As a consequence of this temporary amplifier paralysis, a certain number of detector pulses will be lost in the amplifier, if the detector is a proportional counter, a scintillation detector, or any other detector which has a resolving time shorter than the paralysis time of the amplifier. This type of uncertain loss may be eliminated by the use of dead-time circuits that introduce a dead time that is longer than the longest paralysis time of the amplifier.

Discriminators and Single-Channel Analyzers

A *discriminator* is a circuit that has the function of "sorting" pulses according to their amplitudes. In its simplest form, or in its *integral* mode of operation, it acts to allow only those pulses that are greater than a specified amplitude to pass on to further circuits for other types of analysis.

Since the advent of the integrated circuit (IC), most discriminator designs use a two-input (differential-input) integrated circuit that has either one of two output states, depending on the relative values of the two input voltages. Such an integrated circuit is known as a *differential comparator*.

A simple discriminator circuit that incorporates a differential comparator is shown in Fig. 7-11a. The threshold voltage, V_t, of the discriminator is a dc voltage which is generally selected by the setting of a potentiometer, P. When the amplitude of the input voltage signal, V_i, exceeds the threshold voltage, the output signal (V_0) of the differential comparator changes state for as long as the input is greater than V_t, as illustrated in Fig. 7-11b. This simple circuit is known as an *integral discriminator*, and selects, for further processing, only those input signals that have voltage amplitudes greater than V_t and rejects all those pulses that have amplitudes smaller than V_t.

The integral discriminator is used mostly to discriminate against amplifier noise pulses, while allowing the passage of the larger pulses from radiation detectors. The output pulse is a standard or "logic" pulse, that is, a pulse that has one of two states, or voltages, indicating either a logical "yes" or "1" to signify an input pulse of amplitude greater than V_t, or a logical "no" or "0" for those pulses of smaller amplitude. The output of the discriminator therefore only gives information about the time of arrival of those input pulses that have amplitudes greater than V_t.

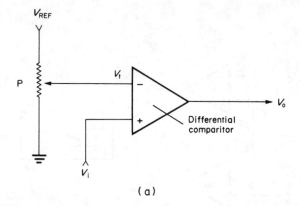

(a)

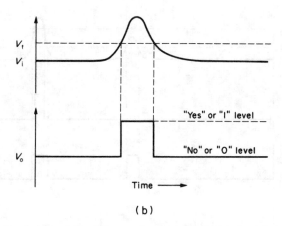

(b)

FIG. 7-11   Simple integral discriminator circuit (a) and (b) relationship between input pulse amplitude V_i, threshold setting V_t, and discriminator output signal V_0. When $V_i \geqslant V_t$, V_0 remains in the "yes" or "1" state.

A single-channel analyzer (SCA) is constructed by combining two discriminator circuits with separately adjustable threshold voltages, as shown in Fig. 7-12a. Additional logic circuitry produces an output pulse only when the voltage amplitude of the input pulse lies between the threshold-voltage settings of the two discriminators. By only passing

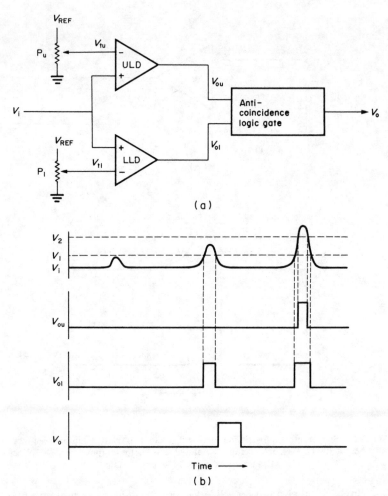

FIG. 7-12 (a) Block diagram of a basic single-channel analyzer; (b) operation of the circuit for pulses below, between and above both the lower V_1 and upper V_2 level discriminator threshold voltage settings. (V_1 and V_2 in (b) are equivalent to V_{tl} and V_{tu} respectively in (a)).

those input pulses with amplitudes greater than a given voltage V_{tl} but smaller than V_{tu}, where $V_{tu} = V_{tl} + \Delta V$, this circuit operates in the *differential* mode. This mode of operation is used when it is necessary to determine the voltage pulse-amplitude, or pulse-height distribution, that is the energy spectrum of the radiations emitted by one or more radionuclides.

As an example, let us consider the γ-ray spectrum of cobalt-60 obtained with a NaI(Tl) detector, such as that shown in Fig. 7-13, and

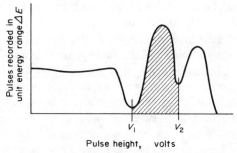

FIG. 7-13   γ-ray spectrum of cobalt-60.

suppose that we are only interested in counting those events that give rise to pulses with amplitudes lying between V_1 and V_2. Referring to the discriminator circuit shown in Fig. 7-12, the lower-level discriminator (LLD) threshold voltage, V_{tl}, is set by P_1 to the voltage V_1, to allow only those pulses with amplitudes greater than V_1 to pass it, such as the second and third pulses shown in Fig. 7-12b. At the same time, the upper-level discriminator (ULD) threshold voltage, V_{tu}, is set by P_u to the voltage V_2, to allow only those pulses with amplitudes greater than V_2 to pass it, such as the third pulse shown in Fig. 7-12b. Referring again to Fig. 7-12b, we note that the first input pulse that is smaller in amplitude than V_1 will pass neither the lower-level nor the upper-level discriminators. The second input pulse, with an amplitude greater than V_1 but less than V_2, will pass the lower-level discriminator to produce an output pulse, but will not pass the upper-level discriminator. The third pulse is, however, larger in amplitude than V_2 and will pass both the lower-level and upper-level discriminators, but will not produce an output pulse from the anticoincidence logic gate, because any ULD output pulse has only one function, namely to close the anticoincidence logic gate to prevent the coincident LLD output pulse from passing through it. This type of logic circuit is called an *anticoincidence circuit*, and produces an output when *one* but *not both* of two inputs are presented, within a predetermined time interval, to the input of the circuit.

In the simplest single-channel-analyzer designs, testing for the anticoincident, or non-coincident, condition (to determine whether the peak amplitude of the input pulse is between V_{tl} and V_{tu}) is delayed until the trailing edge of the input pulse falls below V_{tl}. This ensures that sufficient time has been allowed for both the upper-level and lower-level discriminators to respond and for their respective output states to be tested for the anticoincident condition. If this condition, of non-coincidence, is met, the single-channel analyzer then puts out a *logic pulse*. Such logic pulses are uniform in shape and size.

The voltage difference between V_{tl} and V_{tu} is often referred to as a window, and this type of analyzer is quite suitable for simple counting systems. It will be seen, however, by reference to Figs. 7-11b and 7-12b that the time that the output pulse, V_0, of the differential comparator remains in the "yes" state is dependent upon the amplitude, hence width, of the input pulse, V_i. This simple analyzer can not therefore be satisfactorily used in experiments that require close correlation in time between the SCA input and output pulses. More sophisticated methods, requiring circuits of greater complexity, must be used when time-correlated SCA outputs are required. Such single-channel analyzers are often called *timing-single-channel analyzers*.

Scalers

Most radiation-detector systems give the number of ionizing events occurring within a detector in the form of variable-amplitude pulses; and, in the absence of electronic noise, each pulse corresponds to a detected event. For many practical applications, the only information required is the number of such events occurring in a given interval of time, and this can be provided by means of a simple counting system consisting of a detector and amplifier, a discriminator or single-channel analyzer, and a scaler. The discriminator or SCA is used to select and shape the pulses of interest, and the scaler records and displays the total number of pulses arriving at its input during some selected interval of time.

Modern scalers use a series of integrated-circuit (IC) counters, known as *decade counters*, to register the number of input events (see Malmstat and Enke, 1969). The term scaler is derived from the early use of electronic dividers that scaled down the number of input pulses by known factors, so that the input rates were compatible with the lower recording rates of the electromechanical registers then in use. Although modern scalers are, in reality, counters, the term "scaler" is still used when referring to these devices.

As shown in Fig. 7-14, a scaler of 1 to n decades can be constructed by interconnecting appropriate numbers of decade-counter IC's. Typical commercial counters have six decades. The counting interval is controlled by applying a signal to an input, often referred to as the *count-enable gate*, for a predetermined interval of time. This interval of time is usually controlled by an external or internal timer or clock that can be preset for varying intervals of time. When operating, the scaler registers all input pulses and accumulates them in binary sequences arranged in decades, as shown in the table of Fig. 7-14. This binary-decade format is known as *binary-coded decimal* (BCD). The running content of the scaler is usually made available through an external connector for data acquisition in external devices such as printers and recorders. When a decade counter is at "nine" and receives another input signal the BCD resets to zero and a pulse is transmitted from the overflow output to the input of the next, higher-order, scaler; and so on, for each succeeding stage. A "reset" input is provided to "zero" all the scaler stages at the end of a counting interval, before initiating another.

In order to give a visual display of the accumulated scaler counts, seven-segment light-emitting diodes (LED's), such as those to be found in calculators and digital watches, are most commonly used. (A light-emitting diode consists of a small piece of a semiconducting crystal, such as gallium arsenide, in which electric power is used to raise electrons to the conduction band whence they return to lower-energy levels with the

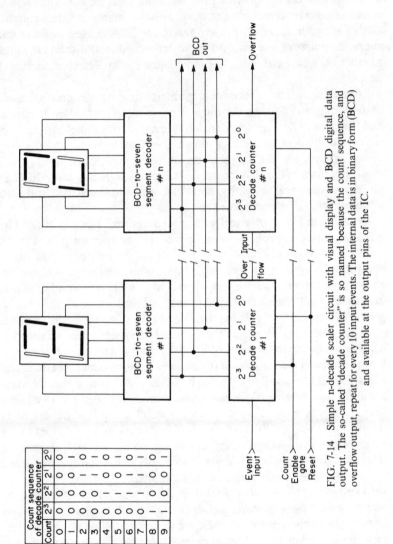

FIG. 7-14 Simple n-decade scaler circuit with visual display and BCD digital data output. The so-called "decade counter" is so named because the count sequence, and overflow output, repeat for every 10 input events. The internal data is in binary form (BCD) and available at the output pins of the IC.

emission of visible radiation.) In order to display the appropriate decimal digit from the BCD output of the scaler IC, the BCD code must be decoded in order that only those segments in the display, that are needed to form the required decimal digit, shall be lit. As indicated in Fig. 7-14, this is accomplished by means of a BCD-to-seven-segment-decoder IC, for each scaler-display pair.

Multichannel Pulse-Height Analyzers

For many types of radiation measurements, an integral discriminator or single-channel analyzer will suffice, but there are other situations in which, because of the limitation of time, neither instrument is adequate. When, for example, one has to obtain the energy spectrum of the radiation emitted in the decay of a short-lived radionuclide, or from a very low-activity sample of a long-lived radionuclide, then it is very desirable to record the entire spectrum simultaneously. This can be accomplished by means of a *multichannel pulse-height analyzer* (MCA). The earliest instruments of this type consisted of, say, n single-channel analyzers composed of n + 1 integral discriminators. Thus a 10-channel or 20-channel analyzer was composed of 11 or 21 integral discriminators, just as a single-channel analyzer comprised two integral discriminators. Each was followed by a scaler, and, as the number of channels was increased, this multi-discriminator type of analyzer became very expensive, both in terms of the electronic components needed to build it and the subsequent maintenance required. There was also the problem of maintaining uniform and stable channel widths. Ideally each channel width, ΔV, should be constant and equal to the same increment in the input-pulse amplitude. (Deviation from this ideal condition is termed *differential non-linearity*.) The disadvantages of using multiple integral discriminators may, however, be overcome in the future as a result of advances in very large-scale circuit integration.

Modern MCA's have 512 to 8192 channels, and use one of several methods of pulse-amplitude analysis which are different from that which has just been described. MCA's having 512 to 1024 channels are usually found to be adequate for use with low-resolution-detector systems, such as those using NaI(Tl) scintillation detectors, whereas MCA's with 4096 to 8192 channels are normally used with high-resolution semiconductor-detector systems.

One of the most widely used methods of pulse-height analysis is based on conversion of pulse heights to proportional intervals of time. This technique was first developed by D. Wilkinson in 1949, and uses just a single, but relatively complex circuit, to determine the height (i.e. the voltage amplitude) of a pulse. When a pulse enters the analyzer, it is first "stretched," or held at its peak amplitude, in order to charge a capacitor to a voltage proportional to the pulse height. Following this, the capacitor is discharged linearly by means of a constant current source (rather than exponentially by means of the familiar RC circuit). The time taken for the capacitor to discharge back to zero voltage is thus directly related to the original pulse height, to determine which it is then only necessary to measure this interval of time. This is accomplished by measuring the number of oscillations of a constant-frequency oscillator that occur during the time of discharge. This entire process is called *analog-to-digital conversion*; "analog" referring to the fact that the charge deposited on the capacitor is proportional to the energy deposited in the detector, and "digital" to the fact that the discharge time of the capacitor is measured digitally by means of a scaler. The time intervals thus determine the relative energies of pulses, and, following each of these time measurements, the content of the scaler is used to "address" (i.e., to select) the location in a computer-type memory that corresponds to the amplitude interval of the input pulse. This selected memory location, or "bin," is incremented by one every time that it is addressed by the analog-to-digital converter (ADC), thus providing a tally of the number of input pulses with amplitudes lying within the voltage interval of the bin. This voltage interval is determined by the ADC input-voltage range and resolution (in number of channels or intervals of time). For example, consider an ADC with a 0 to 10-V range and a resolution of 100 (i.e., capable of sorting the input-pulse amplitudes into bins, or channels, that have increments of 0.1 V). For all input pulses that have an amplitude between 6.5 and 6.6 V the ADC would produce an address of 65 and increase memory location 65 by one, for each such input pulse.

One of the shortcomings of this type of analyzer is the length of time required to convert and *store* pulse. This usually amounts to 10 ns per channel for a typical oscillator frequency of 100 MHz, plus approximately 5 to 10 μs to record the converted pulse in the appropriate memory or channel. Thus, if an analyzer has 8000 channels, the time required to analyze a pulse will vary from 0 to 80 μs, depending on the pulse amplitude, and a further 5 to 10 μs to store it. As a single MCA can analyze only one pulse at a time, during which analysis the MCA input is closed or "dead" and unable to process any additional pulses, the "dead-time" losses become appreciable at what would be considered very nominal counting rates with other instruments. In addition, the variability in magnitude, noted in our example, of these long dead times makes it difficult to correct for dead-time losses in the count-rate data. (See the next section on dead-time losses and circuits.) Without such correction, however, the data would be unrepresentative of the true pulse-height distribution, because more pulses would be lost during the processing of large-amplitude pulses than smaller ones.

This difficulty is eliminated, however, by using special circuitry within the MCA and its constant-frequency oscillator as a "clock," to record only that time during which the MCA is receptive to incoming pulses, and ready to process them. The clock is shut, or "gated," off during the "dead" processing periods, and therefore records only the *live time* of the MCA. MCA's with such "live timers" provide pulse-height-distribution data that are not in error by more than 1 per cent over the normal range of input pulse rates and amplitudes. Other methods of correcting for MCA counting losses can also be used, and may be superior to that of live timing. Such methods are reviewed in section 2.7 of NCRP Report 58.

A live-timed MCA provides a differential pulse-height spectrum of detected ionizing radiations by analyzing large numbers of pulses, and accumulating a histogram, or pulse-amplitude distribution, that is the sum of all pulses sorted into individual bins, or channels, according to their amplitude. Modern MCA's are equipped with a variety of auxiliary circuits for displaying and manipulating the accumulated data. These functions are often provided by fixed circuits, and are termed "hard-wired" functions, but current trends are towards ever increasing utilization of computers (both mini- and micro-computers) as an integral part of the MCA. The use of programmable computers to display and manipulate the data offers a considerable increase in the flexibility of the multichannel analyzer.

Dead-Time Losses and Circuits

For all nuclear pulse-measuring systems, there exists a minimum interval of time that the system, including the detector itself, requires to process a detected event. This interval of time is known as the *dead time*, τ, of the system, and events occurring during this time cannot be processed, and are lost. The fraction of the total events that is lost is a function of the mean event rate, and increases with increasing event rate.

The dead time of the system, depending on its response to events occurring during its dead time, can be classified as one of two types, namely *extendable* or *nonextendable*. Those systems, for which one or more additional events occurring during a dead time, extend the dead time for an additional time τ from the time of occurrence of those events, are classified as having extendable dead times. For example, if another event occurs during a dead time, no further event can be processed until the time interval $(\tau + \Delta\tau)$ has elapsed, where $\Delta\tau$ is the time that has elapsed between the initiation of the first and second events. This time interval $\Delta\tau$ may therefore have any value between 0 and τ. On the other hand, those systems in which events occurring during the system dead time have no effect whatsoever on the dead time, are classified as having nonextendable dead times. For such systems, a new input event can be processed as soon as the dead-time interval τ, following the preceding event, has elapsed, regardless of the number of events occurring during the dead-time interval. Examples of the relationship between input events, randomly distributed in time, and the output pulses from systems with both types of dead time are shown in Fig. 7-15.

The overall dead-time response of most counting systems is usually made up of both extendable and nonextendable components. As we shall see, when we consider count-loss corrections in Chapter 8, it is preferable to have systems with as nearly nonextendable dead times as possible in order to facilitate these corrections. A simple circuit, in use at the National Bureau of Standares for accomplishing this, is shown in Fig. 7-16.

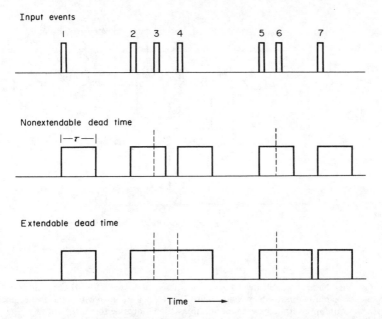

FIG. 7-15 Relationship between input and output events for systems with nonextendable and extendable dead times. For the system with nonextendable dead time τ, events 3 and 6 are lost but events 2, 4, 5 and 7 are recorded because they are separated in time by more than the dead-time interval τ from the last preceding event *to produce an output*. For the system with an extendable dead time, event 4 is also lost because it is separated by less than τ from event 3, which extended the dead-time interval initiated by event 2. For extendable dead times all pulses, that are not separated by at least the interval τ from the preceding event *whether this event is recorded or not*, are lost (after NCRP, 1978).

This circuit uses an integrated circuit (IC) *retriggerable monostable multivibrator* (see, for example, Texas Instruments, 1976), connected to produce an adjustable fixed-width output pulse for each input pulse, that is separated in time from the initiation of a preceding input pulse by more than the fixed time duration of the output pulse. This is accomplished by feeding one of the outputs, $\bar{Q}$, back to the input in such a way as to block the input for the duration of the output pulse. The relationships between the input and output pulses for this circuit are the same as those shown in Fig. 7-15 for nonextendable dead times. The

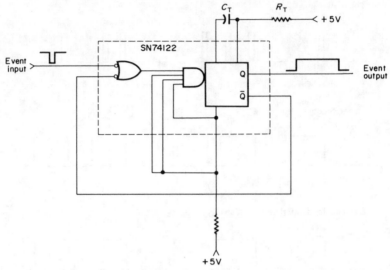

FIG. 7-16   Nonextendable dead-time circuit based on a single SN74122 retriggerable-monostable-multivibrator integrated circuit. The IC itself is shown within the dashed lines with appropriate external connections shown outside the dashed lines.

width of the output pulse, and, hence, the nonextendable dead time τ, is determined by the values of C_T and R_T in the product

$$\tau = 0.45 R_T C_T, \tag{7-8}$$

for C_T in picofarads (and equal to or less than $1000\,\mathrm{pF}$), R_T in kilohms, and τ in seconds. The value of the constant in Eq. 7-8, or even the form of the equation itself, may be different for other types of the IC than that (SN 74122) used in our example. It is common practice to use a variable resistor for R_T to provide a convenient means of varying τ.

Each circuit in a measuring system has an associated dead time, so that, in general, there will be a number of different dead times in series. This makes it impossible to achieve a perfectly nonextendable dead time unless the longest nonextending dead time is placed first in the series (i.e., immediately at the output of the detector), which is usually impractical. In practice, however, when the longest nonextendable dead time is placed later in the series, and the overall dead time is assumed to be nonextendable, acceptable corrections can be made for dead-time losses, provided that these do not exceed about 10 per cent of the count rate.

Section 2.7 of NCRP Report 58 and the references cited therein give a very comprehensive review of the subject of dead-time losses. It also describes how one may determine whether, for a given electronic system, its dead time is extendable or nonextendable.

Electrometers and Direct-Current Measurements

For the measurement of the activities of radionuclides using ionization chambers, it is necessary to measure direct currents ranging from about 10^{-7} to 10^{-13} A. Such currents are typically measured using a high-gain differential amplifier with an extremely high input resistance, of about 10^{14} ohms or more, and small offset currents (differences in current between the two inputs of differential amplifier required to produce zero output) of about 5×10^{-14} A or less. Such amplifiers are known as *electrometer amplifiers*, or simply *electrometers*, and the corresponding low-current measuring circuit as an *electrometer-picoammeter* circuit. These very small ionization currents can also be measured by integrating the current. This is done by collecting charge on a capacitor of known capacitance and measuring the voltage developed in a given time, by means of circuits known as "electrometer current integrators". Low current-measuring instruments based on this principle are known as *electrometer coulombmeters*, or, simply, *coulombmeters*.

Several differently designed electrometers are available (see Keithly Instruments, 1977). These different designs are classified by the kind of input stage used, such as, metal-oxide-semiconductor field-effect transistor (MOS-FET), electrometer tube, vibrating capacitor, varactor bridge, or junction FET. Of these, the vibrating capacitor and MOS-FET are the most commonly used types for present-day measurements of ionization-chamber currents. The vibrating-capacitor electrometer offers the best current and voltage stability and the highest input impedance of any electrometer. Modern MOS-FET designs, however, provide more than adequate performance at lower cost and less complexity for most applications, and are currently becoming the most widely used types of electrometer amplifiers.

All electrometer current-measuring methods rely on converting the input current, from, say, an ionization chamber, to a voltage signal across the input terminals of the electrometer amplifier. The function of the electrometer amplifier is to amplify the corresponding small input voltage signals without loading the signal source by drawing any appreciable current from it. Thus the requirement for very high input resistances (at least several orders of magnitude greater than that of the input source). There are several basic ways in which the current from the signal source can be converted to a voltage proportional to the current at the input of the electrometer. The simplest of these current-to-voltage conversion methods is based on Ohm's Law, the voltage across a shunt resistance, R_s in ohms, being equal to $I_i R_s$, where I_i is the input current in amperes. This method is the basis for the electrometer-picoammeter circuit mentioned earlier. Alternatively the input current can be accumulated on a capacitor producing a voltage equal to Q/C, where Q is the charge accumulated on the capacitor in coulombs, and C is the capacitance of the capacitor in farads. Hence the name *electrometer coulombmeter* or *coulombmeter* for such electrometer-current integrators.

For each of these two basic current-measuring techniques, there are also two applicable possible input circuit configurations. These are known, respectively, as the shunt and feedback configurations. Examples of both are presented in Figs. 7-17 and 7-18 respectively for the picoammeter and coulombmeter measuring systems, together with the corresponding relationships between the input currents and output voltages for each of the four circuits. These relationships have been derived in a text on electrometer measurements (Keithley Instruments, 1977) and further discussed, specifically with regard to ionization-current measurement, in a review by K. Zsdánszky published in 1973. We will, therefore, omit any further detailed discussion of the derivation of these equations or the functioning of these circuits, and refer the interested reader to these two sources of information.

The shunt-circuit configuration (which, incidentally, is also seen to employ feedback) is, however, rarely used for the measurement of very small ionization currents in either the electrometer picoammeter or

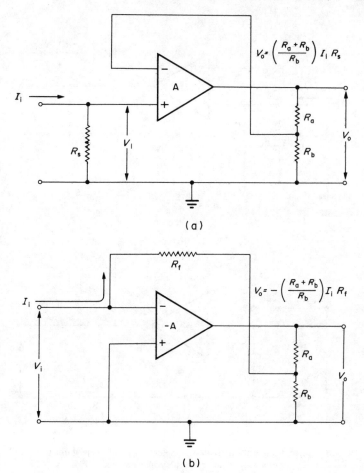

$$V_o = \left(\frac{R_a + R_b}{R_b} \right) I_i \, R_s$$

(a)

$$V_o = - \left(\frac{R_a + R_b}{R_b} \right) I_i \, R_f$$

(b)

FIG. 7-17   Electrometer-picoammeter circuits: (a) shunt-type and (b) feedback-type
(after Keithley Instruments, 1977).

coulombmeter. This is because the negative-feedback configuration
provides much superior performance both in terms of stability and
measurement accuracy. We will, therefore, omit any further reference to
the shunt configurations.

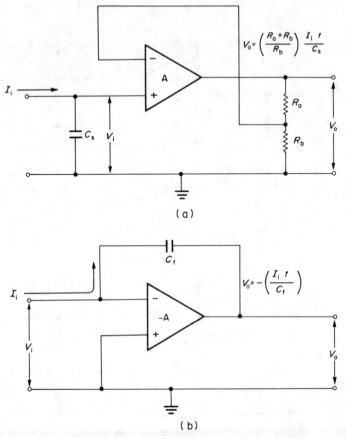

$$V_o = \left(\frac{R_a + R_b}{R_b}\right)\frac{I_i\ t}{C_s}$$

(a)

$$V_o = -\left(\frac{I_i\ t}{C_f}\right)$$

(b)

FIG. 7-18 Current-integration (coulombmeter) circuits: (a) shunt-type and (b) feedback-type (after Keithley Instruments, 1977).

On examination of the relationship between the input current and the output voltage of the picoammeter, as given in Fig. 7-17b, it will be seen that the current sensitivity of the circuit, that is the ratio of V_0 to I_i, is determined by the values of the feedback resistance, R_f, and of the resistances, R_a and R_b. In this circuit the resistor R_f establishes the input voltage of the electrometer, according to $V_i = I_i R_f$, and is referred to as the "ranging resistor." The combination of the resistors R_a and R_b is known as the "multiplier configuration," and their values determine the voltage gain of the electrometer amplifier. Thus,

$$V_0 = -\left(\frac{R_a + R_b}{R_b}\right) V_i \qquad (7\text{-}9)$$

or

$$V_0 = -A_c V_i \qquad (7\text{-}10)$$

could be substituted for the equation given in Fig. 7-17b, where V_i is the input voltage produced and A_c the closed-loop gain of the amplifier as determined by the combination R_a and R_b. Although other variations of this basic electrometer-feedback circuit are possible, this circuit (Fig. 7-17b) provides for several independent means of varying the current sensitivity, namely, by varying the values of R_f, R_a, or R_b. This allows for the design of versatile and specialized electrometer circuits such as that used for "dose calibrators" which is discussed at the end of this section.

In the case of the negative-feedback coulombmeter, shown diagrammatically in Fig. 7-18b, the capacitor in the negative-feedback loop combines both the integrating and feedback functions. For most ionization-chamber measurements of radioactivity, the current output of the ionization chamber, I_i, will be essentially constant during the time of measurement, t. Therefore, the change in voltage across the capacitor due to the accumulated charge would be given by

$$V_c = I_i t / C_f, \qquad (7\text{-}11)$$

where C_f is the capacitance of the integrating capacitor. As mentioned previously, in the section on preamplifiers, an amplifier with negative feedback acts as a sensor and operates to reduce the input voltage signal, across its input terminals, to zero. Thus if $V_i = 0$, then, as $V_c = V_i - V_0$, V_c will equal $-V_0$, and Eq. 7-11 reduces to that given in Fig. 7-18b, namely that I_i is equal to $-V_0 C/t$.

Equation 7-11 shows that the current sensitivity of this circuit can be varied by changing the value of C, the measuring time interval t, or both. More sophisticated circuits can be constructed that use a constant current source to discharge the integrating capacitor at the end of the measurement interval, while timing the discharge of the capacitor by counting a fixed frequency oscillator. This converts the voltage held on the capacitor to a digital signal in a manner analogous to that of the previously described Wilkinson type analog-to-digital converter (ADC). In this, and similar applications, this type of analog-to-digital conversion is referred to as *dual-slope conversion*. The resolution of the ADC is determined by the values of the constant current and the oscillator frequency.

Compared with the electrometer-picoammeter circuit, greater precision and accuracy are possible using current integrating methods. There are two principal reasons for this: (i) by time averaging the input-current signal, short-term noise fluctuations in the input signal average out, and (ii) the collection of charge from large numbers of ionizing events reduces the relative statistical spread of values of the recorded signal about its mean value.

Both the electrometer picoammeters and coulombmeters give output voltage signals that, as shown by the equations in Figs. 7-17 and 7-18, are proportional to the input currents. Thus a strip-chart recorder can provide a permanent record of V_0, or a digital voltmeter can be used to display the value of V_0. The values of V_0, so obtained for samples of a given radionuclide, can then be related to their corresponding activities by calibrating the ionization chamber and its current measuring system by means of standardized samples of that radionuclide.

Other more sophisticated examples of current measuring circuits were described, or referenced, by K. Zsdánszky in 1973.

An ionization-chamber-electrometer combination that is widely used in nuclear medicine is the so-called *dose calibrator*. This instrument, by the use of special circuitry, gives direct readings, in units of the curie, of the activities of radionuclidic sources placed within the reentrant receptacle of the chamber.

A typical dose calibrator might have the feedback-picoammeter circuit shown in Fig. 7-17b, with additional circuitry to provide both a read-out in ranges of different multiples of the curie (e.g., 0 to 1 mCi, 0 to 10 mCi, 0 to 100 mCi, and so on up to 0 to 10 Ci), and the means of adjusting the gain of the electrometer so that the read-out is directly in terms of the measured activity expressed in any one of the foregoing multiples of the curie (e.g., 99.3 mCi, or 0.099 Ci, etc.).

The capability of choosing a read-out in a suitable multiple of the curie is usually provided by a switch, by means of which different values of the resistance R_f can be selected in order to adjust the ratio of V_0 to I_i by factors of 10 (see equation in Fig. 7-17b). Knowing the minimum and maximum values of I_i to be encountered in practice, one can choose a series of values of R_f to divide that range of I_i into four decades, that is from 0 to 10 up to 0 to 10^4 corresponding to 0 to 10 mCi and 0 to 10 Ci.

In order now that a digital voltmeter may read the activity out directly in any chosen multiple of the curie, the electrometer gain, A_c in Eq. 7-6, can be adjusted by changing the value of the resistance R_a. This is normally carried out in conjunction with standards of each radionuclide for which the dose calibrator is likely to be used. Thus if we had a standard of 1.23 mCi of ^{125}I, we could place it in the dose calibrator and adjust R_a until the digital voltmeter read 1.23, in volts, in the 0 to 10 mCi range. This adjustment is normally made by the manufacturer, and the resistor R_a is usually a variable resistor, or one of a group of switch-selectable resistors or interchangeable plug-in resistors.

When the dose calibrator has a variable resistance, R_a, its calibration settings for different radionuclides are normally given by the manufacturer, but it can also be calibrated by the user with appropriate radioactivity standards. In the case of the switch-selectable resistors or plug-in modules each must be identified by the manufacturer for the receptacle of the chamber.

Some dose calibrators now being manufactured can read out in multiples of the curie or becquerel, the choice being made by means of a two-way switch.

References

Ayers, J. B. (1973) Preamplifiers, Chap. 6 in *Instrumentation in Applied Nuclear Chemistry*, J. Krugers (Ed.) (Plenum Press, New York).

Chiang, H. H. (1969) *Basic Nuclear Electronics* (John Wiley & Sons, New York).

Fairstein, E. and Hahn, J. (1965) Nuclear pulse amplifiers—fundamentals and design practice, *Nucleonics 23* [7], 56; *23* [9], 81; *23* [11], 50: *24* [1], 54; *24* [3], 68.

Hatch, K. F. (1973) Amplifiers, Chap. 7 in *Instrumentation in Applied Nuclear Chemistry*, J. Krugers (Ed.) (Plenum Press, New York).

Keithley Instruments (1977) *Electrometer Measurements*, 2nd ed. (Keithley Instruments, Inc., Cleveland).

Littauer, R. (1965) *Pulse Electronics* (McGraw-Hill, New York).

Malmstat, H. V. and Enke, C. G. (1969) *Digital Electronics for Scientists* (W. A. Benjamin, Inc., New York).

Milam, J. K. (1973) Single-channel analysers, Chap. 8 in *Instrumentation in Applied Nuclear Chemistry*, J. Krugers (Ed.) (Plenum Press, New York).

NCRP (1978) See Chapter 6 reference.

Ross, W. A. (1973) Multichannel analyzers, Chap. 9 in *Instrumentation in Applied Nuclear Chemistry*, J. Krugers (Ed.) (Plenum Press, New York).

Sheingold, D. H. (1972) *Analog-Digital Conversion Handbook* (Analog Devices, Inc., Norwood, Mass.).

Texas Instruments (1976) *The TTL Data Book for Design Engineers*, 2nd ed. (Texas Instruments, Inc., Dallas).

Wilkinson, D. (1950) A stable ninety-nine channel pulse amplitude analyzer for slow counting, *Proc. Camb. Phil. Soc. 46*. 508.

Zsdánszky, K. (1973) Precise measurement of small currents, *Nucl. Instr. and Meth., 112*, 299.

8 Radioactivity Measurements

Introduction

BETWEEN 1900 and 1903 one of the very fruitful investigative periods in the history of radioactivity occurred when Ernest Rutherford and Frederick Soddy became collaborators at McGill University. As already described in Chapter 2, two very important papers were published by Rutherford and Soddy, namely "The Cause and Nature of Radioactivity" in 1902, and "Radioactive Change" in 1903. In the latter they stated that the proportional amount of radioactive matter that changes in unit time is a constant, which they called the decay constant, λ. The quantity $-\lambda N_t$ is what we now call the *activity* of a radioactive substance at time t.

Considering the radioactive decay of ^{60}Co (symbolized in Fig. 3-10) the transitions through the two excited levels of ^{60}Ni occur in a very short time of the order of 10^{-12} s or less. (Branching β transitions of the order of respectively 0.1 and 0.01 per cent to the 1.33-MeV and 2.16-MeV levels of ^{60}Ni are not shown in Fig. 3-10.)

No γ-ray transitions are, however, instantaneous but follow the same statistical law, namely that the number of transitions in unit time is proportional to the number of nuclei remaining in a given excited level of energy. As our ability to measure shorter and shorter intervals of time increases, so are these energy levels found to decay with well characterized decay constants, and half lives of the order of nanoseconds and picoseconds.

On the other hand, some of the transitions through excited states of a stable daughter nucleus have half lives of the order of seconds, minutes, and even years, so that excited daughter nuclei may be chemically separated from the parent and may thus be considered to be radioactive in their own right. One of the best known examples of this is that of the β decay of ^{137}Cs where the excited barium daughter nucleus emits 0.662-MeV γ rays with a half life of 2.55 minutes. Such photon transitions with easily observed half lives are called *isomeric transitions*, and the letter "m," for "metastable" is used to designate *isomers* as, for example, ^{137m}Ba, or alternatively, ^{137}Ba^m.

By and large, the number of isomers that exists is limited only by our ability to measure short half lives. In fact all γ-ray transitions are isomeric, but only those whose half lives can be measured are so designated. It is almost a philosophical question as to where the dividing line should be placed. Probably instead of thinking of such transitions from excited states of nuclei as "instantaneous" we should just call them "prompt."

Not all isomeric levels decay, however, by γ-ray emission, or internal conversion. Examples of other modes of decay are: ^{242m}Am, half life 152y with a 0.48-per-cent α-particle branch; ^{99m}Tc, half life 6.02h with a 0.0094-per-cent β-particle branch; and ^{234m}Pa, half life 1.17 min with a 99.87-per-cent β-particle branch.

Because of these developments the concept of the activity of a radionuclide has been broadened from the ideas of the "changing systems" of Rutherford and Soddy or the transforming atoms of later years, to that of nuclear-energy transitions (where mass would also be identified with energy as in Chapter 5). Consequently, while not universally accepted, the following, nevertheless, constitutes a rather precise definition of what is meant when we talk about the quantity "activity":

> The activity of an amount of a nuclide in a specified energy state at a given time is the expectation value, at that time, of the rate of spontaneous nuclear transitions from that energy state.

The *expectation value* is the mean value of a specified statistic of a population that is distributed about that mean value. In the above definition, zero activity corresponds to a stable nuclide. The international SI unit for the quantity "activity" is the *becquerel*, or second to the power of minus one, while the *curie* is accepted as an interim unit of activity, outside the SI system, because of its widespread use.

Radioactivity measurements are made for a great many purposes, probably most frequently to determine the activities of radionuclides and their identities. Other measurements may be concerned with the dispersion of radionuclides in nature, or their behavior in biological or chemical systems, as, for example, in the study of physiological processes and the diagnosis of disease. For such measurements, the procedure of choice will vary with the radionuclide or radionuclides being measured, and the nature of their radiations, and will also depend on the size, composition and nature of the source and the purpose of the measurement itself. As we have seen in the earlier chapters, many types of radiation detectors and measuring systems are available, each with its own strengths and weaknesses for the task at hand. One measure of an experienced and competent investigator is his ability to select those procedures best suited to his measuring objectives within the constraints imposed by the resources available to him.

It is our purpose, in this chapter, to outline briefly the limitations associated with each of the major techniques for measuring the radiations from radioactive substances, with emphasis on those that are applicable to the field of nuclear medicine, and to review those subjects that are relevant to most, if not all, radioactivity measurements, such as statistics, energy resolution, and dead-time losses. In this context, only those procedures that are applicable to establishing either the activity or the identity of a radionuclide will be given primary consideration. Thus, in general, procedures for the measurement of, say, decay-scheme parameters such as half lives and the intensities per decay and energies of emitted radiations will only be described in so far as they are applicable to establishing the identity of a radionuclide.

The physics, or modes of operation, of some commonly used radiation detectors were described in Chapter 6, and the electronic systems used to process the responses from such detectors were described in Chapter 7. Only in certain cases, such as for pulse counting in gas ionization detectors, were the output signals essentially equal to the number of ionizing events occurring within the detector. Even, however, in these detectors output signals could be lost if they occurred too close to a preceding pulse (i.e. within the "dead time" of the system), while in the case of scintillation and semiconductor detectors the overall detection efficiencies may be considerably less than 100 per cent. This efficiency of the detector and its system is often called an intrinsic efficiency. In addition, however, there is a geometrical efficiency arising from the fact that radiations from a radioactive source that is free from scattering are emitted isotropically into a solid angle of 4π steradians, whereas the solid angle subtended by the detector to the source is usually less. Only in the case of the $4\pi\beta$ counters, illustrated in Figs. 6-9, 6-10, and 6-11, do the detector efficiencies approach closely to 100 per cent.

If a detector system has an overall detection efficiency ε (i.e. intrinsic and geometric, including self-absorption, scattering and other attenuation of the radiation from the source), then the activity of a source will be given by the rate that, say, β decays are detected by the system divided by its β efficiency ε_β. Thus if 50 β decays per second are detected by a detector system for which ε_β is equal to 0.5, for a given source-to-detector geometry, then the activity of the source would be $100\,\text{s}^{-1}$. This simple example, however, neglects the corrections that must be made for the background count rate of the detector and dead-time losses in the detector system. Cosmic radiation and radiations from radioactive materials in the environment will be detected, and the average count rate due to this background must be determined in the absence of the source whose activity is to be measured. This background count rate is then subtracted from the average count rate obtained with the source in place.

In general, the efficiency, ε, of a detecting system is unknown, but it can be measured, or the need to know it can be eliminated, by methods to be described later in this chapter. Essentially, there are two general methods by which this can be achieved. In the first, ε is simply measured by determining the count-rate response of the detector system to a calibrated standard source. The ratio of this count rate to the certified activity of the source gives the efficiency of the detector system. The activity of an unknown source of the same radionuclide may be determined by measuring the count rate for the unknown source when it is substituted for the standard, in as identically the same geometry as possible and taking care not to change any of the parameters (or, hence, the geometry) of the detecting system. The activity of the unknown source is then equal to the ratio of the mean of the count rates recorded for it to that for the standard, multiplied by the certified activity of the standard. Alternatively, the activity of the unknown is equal to its mean count rate divided by ε. The second method applies only to sources of radionuclides that simultaneously emit two or more types of radiation that can be observed separately with different efficiencies. Count-rate equations can then be established to give the activity of such sources directly. Such methods will be described in the subsequent sections on *coincidence counting*.

From the above it is apparent that two types of activity measurements can be made, namely in terms of calibrated standards, or independently of any such standards. The former are termed *indirect, relative*, or *comparative* measurements, while the latter are characterized as *direct, fundamental*, and sometimes, *absolute*. As noted, however, in Chapter 2, radioactivity is a random phenomenon and the decay rate is proportional to time and to the number of radioactive atoms present at a given time, and the fraction of the number of atoms decaying in a given interval of time fluctuates statistically. Activities can therefore only be expressed in terms of *mean values*, which are often also referred to as *expectation values*. There is implicit statistical uncertainty in any measurement of activity, so that the term "absolute," in its usual connotation, scarcely applies.

The γ-ray spectrum of ^{60}Co shown in Fig. 7-13 is another example of the random or statistical nature of the energy exchanges between the incident radiation and the detector, in this case a NaI(Tl) crystal. This spectrum is typical of that which would be obtained with a multichannel analyzer which collects into a given channel pulses of the same amplitude. We know that the pulse amplitude, or pulse height, is proportional to the number of photoelectrons released from the cathode of the phototube, and that the incident γ ray has a very sharply defined energy. But in the process of converting this energy first into photons in the crystal, and then these photons into photoelectrons in the phototube and in the subsequent electron multiplication, statistical fluctuations occur around the mean energy of the photopeak as recorded by the multichannel analyzer.

The statistics of radioactive decay and of peak energy resolution will be discussed in the next two sections.

Statistics

Radioactivity is a random, or stochastic, process in time. As noted in Chapyter 2, $\lambda N_t \, dt$ represents the probability that dN_t atoms will decay in the interval of time dt at time t. Also, by integration, we obtained the radioactive decay law of Rutherford and Soddy, namely if there are N_0 radioactive atoms at time $t = 0$, then the number at time t will be N_t where $N_t = N_0 e^{-\lambda t}$. But because radioactivity is a random process, $e^{-\lambda t}$ actually represents the probability that an atom will not decay in time t, and, therefore, $1 - e^{-\lambda t}$ represents the probability that it will decay in time t. Thus the expected number of atoms decaying in time t will be $N_0(1 - e^{-\lambda t})$. This is also called the *expectation value*.

The statistics of radioactive decay are rather analogous to those of tossing a coin. In the former an atom will or will not decay in a given interval of time, and in the latter the coin will or will not be heads. The possible experimental outcomes are expressed in discrete whole numbers. In other words 101, 95, 94, 104, and 99 atoms may be observed to decay in five consecutive observations each lasting for time t; while in three separate experiments in each of which a coin is tossed 100 times we might throw heads, say, 54, 47, and 50 times.

Suppose we have some radioactive material consisting of a large number N_0 of atoms such that the number, $n(= N_0 - N_t)$, decaying in a time t does not significantly alter N_0. If we now carry out a large number of observations of n, each lasting for time t, we would observe that the values of n are integers distributed over the range from 0 through N_0. The probability P_n of observing exactly n decays in time t in any given experiment can be expressed, in terms of N_0, n and the probability $w(= 1 - e^{-\lambda t})$ of a single atom decaying in time t, by the binomial probability distribution. In practice it is usual to use either the Poisson distribution or the normal distribution, to approximate the binomial distribution.

The Poisson distribution is derived from the binomial distribution when N_0 is very large, and n is very small compared with N_0, from which it follows that the measurement time t must be much less than the half life, $T_{1/2}$. The Poisson distribution therefore approximates closely to the conditions of radioactive decay where N_0 is generally very large, and n is very small compared with N_0. The derivations of the binomial and Poisson distributions are given in many texts, most recently in NCRP Report 58, and will not be repeated here. The normal distribution, the derivation of which is also given in this Report, is an approximation to the Poisson distribution for n large (100 or more), but still small compared with N_0 (which is of the order of 10^{16} atoms for 1 μCi of ^{14}C).

The normal approximation to P_n is, however, a continuous function of n where the other two are discrete and only give values of P_n for integral values of n. The following sections will briefly describe these distributions.

The Binomial or Bernoulli* Distribution

This distribution gives the probability P_n of observing exactly n decays of N_0 atoms in time t to be

$$P_n = \frac{N_0!}{n!(N_0 - n)!} w^n (1 - w)^{N_0 - n}, \tag{8-1}$$

where $w = 1 - e^{-\lambda t}$. As implied in the last paragraph, the binomial distribution is generally used *only* for small values of N_0; otherwise $N_0!$ becomes inordinately large.

In Fig. 8-1 the binomial distribution is plotted as a function of n for N_0 equal to 50, but for two different values of w, namely 0.1 and 0.5. As would be expected, the binomial distribution for w equal to 0.5 is symmetrical around the mean, or expectation value. These examples are not, however, relevant to the practical applications of radioactivity as N_0 is, here, very small, compared with the numbers of atoms of a radioactive substance, and w is rather large. Considering the conditions that normally apply to radioactivity, N_0 is a very large number, of the order of 10^{16} or more, n is very small compared with N_0 and, hence, w is very small so that λt is also very small compared with unity (the last condition also being equivalent to saying that the measurement time t is very small compared with $T_{1/2}$, because $\lambda = 0.693/T_{1/2}$). Under these conditions, the binomial distribution is closely approximated by the distribution derived by S. D. Poisson.

The Poisson Distribution

The Poisson distribution for the probability, P_n, of observing n decays in time t is

$$P_n = \frac{(wN_0)^n e^{-wN_0}}{n!}, \tag{8-2}$$

* Jacques Bernoulli, 1654–1705.

where $wN_0 (= N_0(1 - e^{-\lambda t}))$, is the expectation value of the number of decays in time t for a radionuclide having a decay constant λ. Thus the total number of radioactive atoms, N_0, which is a very large number, no longer appears alone in the distribution function, but only as the product of wN_0 which is approximated, in practice, by $\bar{n}$. As before, $\bar{n}$ is the average value of a large number of experimental observations of the number of decays in a given time, which, again in practice, is of the order of magnitude that can be handled by modern laboratory equipment in reasonable intervals of time.

By making a number of observations of n, an observed average value, $\bar{n}$, can be calculated and substituted for the theoretical mean value, wN_0, in Eq. 8-2 to obtain a plot of the appropriate Poisson distribution. The larger the number of observations, the closer will the observed average value approach to the theoretical mean of the population, and the better should be the Poisson fit.

While the Poisson distribution is an adequate approximation to the statistics of radioactive decay, it is also used to represent random phenomena in everyday life. One of its earliest applications was to represent the distribution, from year to year, of the numbers of troopers in the Prussian cavalry that had been kicked to death by their horses.

In Fig. 8-2 plots of the Poisson distribution are shown for wN_0 equal to 50 and to 100. In Fig. 8-1 the Poisson distribution is also plotted for wN_0 equal to 25 in order to compare it with the binomial distribution for N_0 equal to 50 and w equal to 0.5. The value of N_0 is not yet sufficiently large, however, compared with $\bar{n}$ for there to be very close agreement between the two distributions, but, as N_0 increases relative to $\bar{n}$, the two distributions will merge into one. It will be noted from these figures that the Poisson distribution is closely symmetrical around the mean value of n, and that P_{n-1} is equal to P_n for n equal to wN_0.

Deviations from the Mean Value

As is clear from Figs. 8-1 and 8-2, and also from the stochastic nature of the decay process itself, there will be a considerable spread in the values obtained for the number of decays per unit time in a series of experiments designed to measure the activity of a source. It is therefore of great importance to obtain a measure of that spread, when forming an

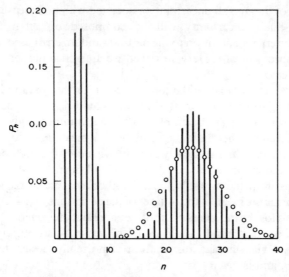

FIG. 8-1   Binomial distributions (vertical bars) for $N_0 = 50$ and $w = 0.1$ and 0.5. Poisson
distribution (circles) for $wN_0 = 25$.

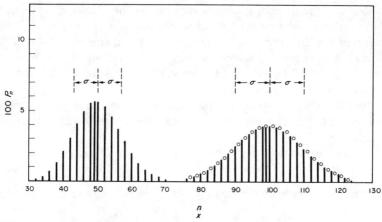

FIG. 8-2   Poisson distributions (vertical bars) for $wN_0 = 50$ and 100. Normal
distribution (circles) for $\mu = \sigma^2 = 100$.

estimate of the overall uncertainty associated with any measurement of activity. Such uncertainty will be composed of, often unknown, systematic error in addition to the natural random error associated with the measurement of n. Here, however, we are concerned only with the random error.

One way to summarize the random error, or the spread of observed values of n_i around the average value $\bar{n}$, would be to list the values of $(n_i - \bar{n})$. It is, however, preferable to have one number to give a sense of the magnitude of the random spread, but the mean deviation, the mean value of all $(n_i - \bar{n})$'s is clearly zero by virtue of the method used to calculate the mean, but use could be made of the mean of the absolute deviations from the mean, $|n_i - \bar{n}|$. Another measure of the scatter, or dispersion, is the mean of the squares of the deviations, $(n_i - \bar{n})^2$, and this is the basis for the measures of the dispersion that are generally accepted, namely the *variance* and the *standard deviation* of a distribution. The true standard deviation of a distribution is normally designated by the symbol σ, and is given by

$$\sigma = \sqrt{\dfrac{\sum\limits_{i=1}^{k} (n_i - \mu)^2}{k}}, \tag{8-3}$$

where n_i is the number of events recorded in the ith of k independent observations (with k conceptually approaching infinity), and μ is the true mean of the distribution. The variance is the square of the standard deviation. It is usually denoted simply by σ^2, and sometimes, but not usually, by "var." In real life it is impossible to measure μ, which is only equal to $\bar{n}$ when k approaches infinity.

It is possible to obtain an estimate of σ by substituting the average value $\bar{n}$, obtained from the k experimental observations, as an estimate of μ into Eq. 8-3.

If the k observations are independent, we define this as constituting k *degrees of freedom*. This means that the k observations are not correlated in any way. But, in calculating the sum $\sum\limits_{i=1}^{k} n_i/k$, we have introduced a correlation between the deviations $n_i - \bar{n}$ in that $\sum\limits_{i=1}^{k} (n_i - \bar{n}) = 0$, and we are therefore left with only $k - 1$ degrees of freedom. Accordingly, we form an estimate s of σ, such that the estimated variance s^2 is given by

$$s^2 = \frac{\sum_{i=1}^{k} (n_i - \bar{n})^2}{k - 1}. \tag{8-4}$$

For a more detailed treatment of this subject, NCRP Report 58 should be consulted. In particular, Table 18 of this Report gives examples of several population parameters and their estimators.

The standard deviation in the case of the Poisson distribution can be calculated to be *numerically* equal to the square root of the true mean value, wN_0, of the distribution (see Appendix B of NCRP Report 58). Thus the theoretical variance is equal to the true mean, and an estimate of the variance is given by $\bar{n}$. The estimated standard deviation is a very useful indication of the magnitudes of the deviations from the average value, $\bar{n}$, that will have been observed in a series of measurements of the activity of a radioactive source, and its units are the same as those of the experimental data (i.e. becquerel, curie, or second to the power of minus one).

If the average number, $\bar{n}$, of the radioactive decays observed in a given interval of time, in a series of separate experiments, is large compared with unity, but is still small compared with N_0, and if we consider a range of values of a variable x around the true mean value μ so that $x = (n - \mu)$, the absolute value of x being much less than μ, then the Poisson distribution can be approximated by the *normal distribution*, or *normal error law*.

The Normal Distribution or Error Law

The normal distribution was apparently first deduced by A. De Moivre in about 1721, then by P. S. Laplace in 1774, and by K. F. Gauss in 1809 (see footnote 8 of W. E. Deming and R. T. Birge, 1921). It is nevertheless credited to Gauss and is now very frequently referred to as the Gaussian distribution. It is also often called an "error law" because, as with the earlier binomial and later Poisson distributions, it attempts to describe the deviations of the values of a statistic from the "true" value that is generally assumed to be represented by the mean. Actually, in radioactivity it is almost fruitless to think in terms of a true value. From a large number of observations we can calculate an average value that will have a higher probability of being observed than the rest, but radioactivity is a random, or stochastic, process in which one expects "truly" to observe disintegration rates that are greater and less than the average value.

In terms of the variable x, where x equals $n - \mu$, the form of the normal distribution is

$$P_x = \frac{1}{\sqrt{2\pi\sigma^2}}e^{-x^2/2\sigma^2}. \qquad (8\text{-}5)$$

The normal error curve is therefore a function of both x and of the standard deviation σ.

The normal distribution can, however, be derived as an approximation to the Poisson distribution when $|x| \ll wN_0$ (see Appendix B of NCRP Report 58), and Eq. 8-5 can then be written in the form

$$P_x = \frac{1}{\sqrt{2\pi\bar{n}}}e^{-x^2/2\bar{n}}, \qquad (8\text{-}6)$$

in which, as an approximation, $\bar{n}$ is substituted for the true variance σ^2 ($= wN_0$ for the Poisson distribution), and x is substituted for $n - \mu$. The normal curve gives P_x as a *continuous* function of x, as distinct from the discrete integral steps of the binomial and Poisson distributions. The normal distribution is plotted in Fig. 8-2 for the case where $\sigma^2 = \mu = 100$, and this may be compared directly with the plot of the Poisson distribution for the same value of 100 used for the expectation value wN_0, or $N_0(1 - e^{-\lambda t})$.

Confidence Limits

The probability that any given observation n lies between $+\infty$ and $-\infty$ is equal to unity. But what is the probability, $P(v, -v)$, of its lying between, say, values $+v$ and $-v$ on either side of the mean value of a distribution? The Poisson distribution is that which is generally used in the processing of radioactivity data, but, for large values of $\bar{n}$ that are still small compared with N_0, and for $x \ll \bar{n}$, the Poisson distribution is closely approximated by the normal error law. We have seen, moreover, that the Poisson distribution (Eq. 8-2) applies only to discrete integral values of n, whereas the normal error law (Eq. 8-5) is a continuous mathematical function susceptible to ordinary mathematical manipulation. An answer to the above question may therefore be obtained by considering the probability of an observation lying between x and $x + dx$, as given by the normal error law, and then integrating between the limits $+v$ and $-v$. Thus

$$P_{(v, -v)} = \frac{1}{\sqrt{2\pi\sigma^2}} \int_{-v}^{+v} e^{-x^2/2\sigma^2}\, dx,$$

or
$$P_{(v, -v)} = \frac{2}{\sqrt{2\pi\sigma^2}} \int_{0.}^{+v} e^{-x^2/2\sigma^2}\, dx. \tag{8-7}$$

Values of the integral in Eq. 8-7 can be obtained from tables of the normal distribution, as found, for example, in National Bureau of Standards Handbook 91, but it can also be evaluated by integrating, term by term, the converging series for e^{-h^2}, namely

$$e^{-h^2} = 1 - h^2 + \frac{h^4}{2!} - \frac{h^6}{3!} + \cdots, \tag{8-8}$$

where $h^2 = x^2/2\sigma^2$.

By this process, we obtain

$$P_{(v, -v)} = \sqrt{\frac{2}{\pi}} \left\{ v/\sigma - \frac{1}{3} \frac{(v/\sigma)^3}{2} + \frac{1}{5} \frac{(v/\sigma)^5}{2!2^2} \right.$$

$$\left. - \frac{1}{7} \frac{(v/\sigma)^7}{3!2^3} + \cdots \right\}. \tag{8-9}$$

Substitution of 1, 2 and 3 for v/σ in Eq. 8-9 gives $P_{(v, -v)}$ respectively equal to 0.683, 0.954 and 0.997. In other words, if a long series of experiments is carried out for equal intervals of time, 68.3 per cent of the observations should give values of n lying within the limits $+\sigma$ and $-\sigma$ from the mean, $\bar{n}$, 95.4 per cent lying within the limits $+2\sigma$ and -2σ, and 99.7 per cent lying within the limits $+3\sigma$ and -3σ. The greater the number of observations, or experimental determinations, of n, the closer will the values obtained for n conform to this distribution.

By the method of successive approximation, it can also be shown by substitution in Eq. 8-9 that for $v/\sigma = 0.6745$, $P_{(v, -v)} = 0.500$. This signifies that 50 per cent of the observations on the normal error curve will lie within the limits $\pm 0.6745\sigma$ and 50 per cent will lie outside. This quantity, 0.6745σ, is known as the *probable error*.

These results are summarized in the following table, for the normal distribution of a large number of observations:

TABLE 8.1

Limits from the mean value	$\pm 0.6745\sigma$	$\pm\sigma$	$\pm 2\sigma$	$\pm 3\sigma$
Confidence or probability that values will lie within these limits	0.500	0.683	0.954	0.997

As in this discussion, probabilities are usually given as fractions or as percentages. It is emphasized, however, that only for large numbers of observations will the data begin to conform to the Poisson or normal distributions and that in all practical cases the estimated value $\bar{n}$ and estimated standard deviation, s, will be only approximations to the true, but never known, mean value, μ, and standard deviation, σ.

The derivation of all the preceding statistical relations was based on the fact that radioactivity is a random, or stochastic, phenomenon and that observations of the numbers of radioactive transitions in equal intervals of time can be treated statistically. Normally, however, we do not measure radioactive transitions but observe counts, or count rates, using radiation-detecting systems that often have detection efficiencies, ε, less than unity. In this event the statistical analysis will still apply provided ε remains constant, but will be scaled down by ε, and the expectation value of the number of counts in time t will be $\varepsilon w N_0$ or $\varepsilon N_0(1 - e^{-\lambda t})$.

The estimate of the standard deviation, s, derived from Eq. 8-4, is for a set of k individual observations comprising an empirical distribution. If we make k observations each of duration Δt in which a mean value of 10^6 counts due to a radioactive source is observed, then the standard deviation would be approximately 10^3 counts or approximately 0.1 per cent. In other words, about 99.7 per cent of the observations would give values of n lying between 997×10^3 and 1003×10^3 (i.e., $\bar{n} \pm 3s$). The greater the number of observations, the closer would we conform to this distribution. If, on the other hand, $\bar{n}$ was only 100 in time Δt, then the estimated standard deviation would be 10, or 10 per cent, and, provided k was reasonably large, some 99.7 per cent of the k observations would give values of n lying between 70 and 130. Therefore, it is important not only that k shall be large in order to verify the distribution, but that the counting time Δt shall be as long as practical in order that each individual value of n shall be as large as possible. In this way we obtain a smaller fractional width, $s/\bar{n}$, to the statistical distribution. In practice, if one is only interested in establishing the mean of the distribution one obtains as large a number of counts as possible to obtain "good statistics".

Suppose, however, that we did have time to make large numbers of observations, and that the half life of the radioactive source was long enough for this to be done. Then we could collect many sets of k observations with mean values of $\bar{n}_1, \bar{n}_2, \bar{n}_3$ etc. In this event, the values of the $\bar{n}$'s would be expected to conform more closely to a normal distribution and the standard deviation of these mean values would be expected to be smaller than the standard deviations of individual values of n. It is in fact possible to estimate the standard deviation of a series of mean values, known as the *standard error*, from a single set of k observations, but it is beyond the scope of this short book to give the derivation. Those interested should consult any text on statistics, such as that by L. G. Parratt. Suffice it to say here that the estimate s_m of the standard error can be expressed in terms of the estimate of the standard deviation, s, of a set of k observations by the simple relation

$$s_m = s/\sqrt{k}$$

Combining this result with Eq. 8.4 gives

$$s_m = \sqrt{\frac{\sum\limits_{i=1}^{k} (n_i - \bar{n})^2}{k(k-1)}} \qquad (8\text{-}10)$$

Once again, the estimate of s_m is better the larger the number, k, of observations. This matter was investigated by W. S. Gosset, in 1908, who, using the pseudonym "Student," derived so-called Student t-factors which can be found tabulated in statistical treatises as functions of the number of independent observations and the desired confidence intervals within which the true mean might be expected to lie. Such a table of t values is given in NCRP Report 58 and in the International Commission of Radiation Units and Measurements Report 12. This latter report deals specifically with the treatment of random and systematic error in the certification of radioactivity standards. The value of s_m given by Eq. 8-10 must be multiplied by the appropriate t factor in order to obtain the required confidence limits for the true mean.

Thus, as an example, if the number of degrees of freedom, k − 1, is equal to 10 then s_m must be multiplied by t = 3.169, in order to obtain a confidence interval of 99 per cent for the mean. Another way to phrase this is to say that for a true mean value μ, we would expect 99 per cent of all future experimental average values $\bar{n}_2, \bar{n}_3$, etc., obtained from future sets of 11 independent observations to lie within $\pm 3.169 s_m$ of μ. At either extreme, the t factor for the 99-per-cent confidence level is equal to 63.657 for k as small as 2, and approaches 2.576 as k approaches infinity.

Propagation of Error

This topic refers to cases where, say, a radioactivity measurement may involve two or more mathematical processes such as addition, subtraction, multiplication or division of estimated mean values and their estimated standard deviations. How does one combine the random errors of two or more distributions into one statement of error? An example that is most commonly encountered in everyday work is that of subtracting the background count rate of a detecting system from the mean count rate observed with a radioactive source. The answer is simple, namely that both in the addition and subtraction of two independent randomly distributed variates, the variance of the combined error is equal to the sum of the variances of the separate errors. Thus if $\bar{r}_{s+b}$, and s_{s+b} are the estimated mean count rate and standard deviation of the count rate, obtained for a radioactive source, and $\bar{r}_b$ and

s_b the corresponding quantities for the background, then $\bar{r}_s$ and s_s will be given by

$$\bar{r}_s = \bar{r}_{s+b} - \bar{r}_b,$$

and

$$s_s^2 = s_{s+b}^2 + s_b^2,$$

provided that $\bar{r}_s$ and $\bar{r}_{s+b}$ are statistically independent. This process of obtaining s_s from the square root of the sum of the squares of the individual estimates of the standard deviations is often referred to as *addition in quadrature*. The values of s are derived from the *total number* of counts obtained in each measurement of $\bar{r}$. Thus, if t is the counting period for any $\bar{r}$, then s per unit time is $(\bar{r}t)^{1/2}/t$, or $(\bar{r}/t)^{1/2}$.

Many other propagation-of-error formulae are given in Table 21 of NCRP Report 58, which should be consulted for more complicated operations than addition or subtraction.

The preceding paragraphs refer to the propagation of random error but systematic errors are always present in any measurement. In the case of the latter an experimenter usually tries to estimate the maximum value of conceivable systematic error in the various operations associated with a measurement, such as weighing, dilution and counting, and then adds them together linearly to give an estimate of total systematic error. By the term "conceivable," we mean the errors that the experimenter can think of and assess! Such errors are not normally susceptible to probabilistic treatment although some investigators, notably and persuasively S. Wagner and J. W. Müller, have adduced interesting reasons for so treating them. W. J. Youden used to make a slight concession to probability by saying that if one had sources of not less than five or six systematic errors that were about equal in magnitude, it was unlikely that all would be of the same sign. Thus if there were six possible systematic errors of about 1 per cent each, two and four of them might be of the opposite sign and cancel; thus one might be justified in taking one-half or one-third of the total of the absolute values of the estimates, for not less than five or six possible sources of systematic error.

One further point for consideration is how one should treat the stated random and systematic components of error of a radioactivity standard that is used to calibrate equipment, that, in turn, is to be used to measure the activities of other sources. The stern answer is that *both* the random and systematic components of error of the standard should be treated as systematic and added linearly to the errors, random and systematic, associated with the use of the equipment to calibrate other sources. The reason is that although the random uncertainty of the standard is stated as "plus or minus" a given percentage of the estimated mean value, it is strictly *either* plus *or* minus (it cannot be both!), and is therefore passed on to a user as a systematic error.

In this context it is perhaps desirable for a standardizing laboratory to match its statements of the estimated limits of uncertainty. That is, if the 99-per-cent confidence limit is quoted for the random error, the maximum conceivable systematic error should be given; if the 50-per-cent confidence limit is taken, it is perhaps reasonable to quote half the maximum conceivable systematic error. In either case, and in *all* cases, the methods used to estimate the uncertainties should be fully explained on the certificate, accompanying a radioactivity standard, so that they may be unscrambled and remanipulated to suit the needs of the user.

Goodness of Fit

If we have a set of k observations of n counts due to a radioactive source, in a fixed interval of time (i.e. the count rate due to the source), how can we determine whether the limited number of data points obtained fits a Poisson distribution? One way is to estimate the mean from the experimental data and then to use this value in either Eq. 8-2 or Eq. 8-6 to plot the appropriate Poisson histogram or normal curve. The experimental data can then be entered on to either plot and compared directly with them. F. Galton, in 1899, suggested the use of "normal-probability" graph paper in which one of the ordinates was adjusted to give a linear plot if the average value of groups of adjacent normally distributed observations were plotted directly on it (see W. J. Dixon and F. J. Massey, 1969).

A rigorous test for normality of data was devised by Karl Pearson in 1900 based on mean-square deviations between the observed and expected values in different subdivided groups of the experimental data. This is called the χ^2, or chi-squared, test. An earlier, simpler and less sensitive but somewhat similar test had been suggested by W. Lexis in 1877.

In the Pearson chi-squared test a set of observations is assembled in ascending order of magnitude, from large negative to large positive about the mean, for checking against the normal distribution. They should then be subdivided into at least five or ten consecutive groups. An estimated mean value and standard deviation of the whole distribution is calculated from the data, and then the expected frequency $E(v_j)$ is calculated for each group using P_n summed or integrated for that subdivision. $E(v_j)$ should be at least five for each group. Chi-squared is defined in terms of the v_j observations in each of the J groups as

$$\chi^2 = \sum_{j=1}^{J} \frac{(v_j - E(v_j))^2}{E(v_j)}, \qquad (8\text{-}11)$$

where $E(v_j)$ is the expected, or expectation, value of v_j calculated from the assumed distribution, in this case the normal distribution. From the value of χ^2 so calculated and the number of degrees of freedom, the probability p of χ^2 exceeding the value derived from Eq. 8-11 can be obtained from tables or standard curves giving the relationship between these three quantities. The number of degrees of freedom is equal to $J - 1$, less the number of parameters, such as μ and σ, that must be estimated from the data. In this case there would be $J - 3$ degrees of freedom. If the value of p is not greater than 0.9 nor less than 0.1, the assumed distribution has a high probability of fitting the observed data. Lower or higher probabilities of χ^2 exceeding the value calculated from Eq. 8-11 can be equally suspicious. If χ^2 itself is calculated to be zero, from Eq. 8-11, then this would mean the fit was perfect which should also be viewed with considerable suspicion.

For a more detailed treatment of the subject, reference should be made to the textbooks of Robley Evans and L. G. Parratt. The former also gives very interesting and instructive examples of the application of the chi-squared test to measurements of radioactivity.

The Lexis test of fit was somewhat similar to the chi-squared test in that he considered the ratio, L^2, of the average mean-square deviations (around the estimated mean value) to the estimated mean value calculated from *all* k observations, without dividing them into groups.

Thus

$$L^2 = \frac{\dfrac{1}{k}\sum_{i=1}^{k}(n_i - \bar{n})^2}{\bar{n}}. \qquad (8\text{-}12)$$

The numerator of this equation is a very close approximation, for large values of k, to the estimated variance, s^2, given in Eq. 8-4, and, for a Poisson distribution, or for the normal distribution as derived from the Poisson distribution, s^2 is very closely equal to $\bar{n}$. Thus, when applied to measurements of radioactivity, Eq. 8-12 can be expected to give

$$L^2 \approx s^2/\bar{n} \approx 1.$$

Thus the Lexis criterion reduces to a statement that if the observed count-rate data conform to a Poisson distribution, the estimated variance s^2, calculated from the data using Eq. 8-4, should approach more and more closely to equality to $\bar{n}$ as k becomes large.

Robley Evans cites an interesting experiment carried out by L. F. Curtiss in 1930 to verify that α-particle count-rate data conformed with the Poisson distribution. The α-particle source was ^{210}Po, a material well known to "creep" by virtue of its recoiling molecular aggregates. The Lexis criterion was used in order to establish when the source became sufficiently stable to enable statistically significant data to be collected. As the source aged, the "creeping" diminished, and L^2 approached more and more closely to unity.

Rejection of Data

The question often arises "what does one do about 'outliers' in a series of observations?" This question may not arise in the use of a dose calibrator in a nuclear medicine laboratory because the read-out represents the integrated response to a large number of electric charges, and, although the reading will fluctuate about an average value, the fluctuations will not be large. But what should one do if one of, say, five count-rate measurements is very different from the average value of the five measurements? The measurement is not necessarily a wrong one, because, in observations of a stochastic process, the results obtained will be distributed widely about the mean, and a very divergent value may be amongst the first obtained. There is thus always a danger of either rejecting a good measurement, or of not rejecting a bad one.

Some experimenters use an arbitrary criterion and, having calculated an estimate, s, of the standard deviation from Eq. 8-4, they reject any value that is different from the average value by more than, say, $3s$ or $4s$. Over 80 years ago W. Chauvenet developed a criterion that also took into account the number of observations, k, in a given group of measurements, by stipulating that if any observation had a probability of being observed of less than $1/2k$ it could be rejected. In other words, if, as above, k = 5 then an observation having a probability of being observed of less than 0.10, or 10 per cent, could be rejected; or it should not be rejected if it had a probability of being observed of more than 90 per cent. One can then substitute this probability of 90 per cent into the left-hand side of Eq. 8-7 to calculate the rejection limits $\pm v/s$.

Chauvenet's criterion is, however, no longer generally acceptable because, if the sample is a small one, only very approximate values of μ and σ may be available, and its use has also been found to result in the rejection of too many good data.

T. A. Willke wrote a short note in 1965 entitled "Useful Alternatives to Chauvenet's Rule for Rejection of Measurement Data," in which the alternative methods cited are based on calculated probabilities of rejected data being good data. First a series of k observations $x_1, x_2, \ldots, x_k$ is arranged in ascending (or descending) order of magnitude, so that $x_1 \leqslant x_2 \leqslant \cdots \leqslant x_k$. As the true population mean, μ, is generally unknown, the sample mean, or average value, $\bar{x}$, is then used to calculate the largest deviation, Δ from the average value, Δ being the larger value of $|x_1 - \bar{x}|$ and $|x_k - \bar{x}|$. The estimated standard deviation is then calculated from the sample, or group, of observations and the ratio $C = \Delta/s$ is calculated. Tables have been prepared, by C. P. Quesenberry and H. A. David, that, for different values of k, give the values for C that correspond to the 1-per-cent and 5-per-cent probabilities of rejecting a good observation. Thus for k = 20, C = 2.999 for a 1-per-cent risk of rejecting a good observation (i.e. Δ = 2.999s), and C = 2.707 for a 5-per-cent risk of rejecting a good observation (i.e. Δ = 2.707s); for k = 10, the respective values of C are 2.481 and 2.289.

W. J. Dixon and F. J. Massey in their *Introduction to Statistical Analysis* also give tables of estimates for a range of probabilities between 0.5 and 30 per cent of rejecting good data, for different values of k in terms of $(x_2 - x_1)/(x_k - x_1)$, and $(x_3 - x_1)/(x_{k-2} - x_1)$.

It should be emphasized that all of the methods discussed above should be treated as *guides* rather than *rules*.

Energy Resolution of Detector Systems

The output-pulse amplitudes, or pulse heights, of most of the detecting systems that we have considered, have been proportional to the amounts of radiation energy deposited within their sensitive volumes. Statistical fluctuations can, however, be introduced between all stages of the detecting system. These include, for example, (i) the conversion of radiation energy to electron-ion or electron-hole pairs, Compton recoil electrons, photons, and, in solid-state detectors, trapped electrons, (ii) transfer of the detector output signal to a phototube, preamplifier or amplifier and, (iii) transfer to the sorting and recording, or final read-out, stage.

Thus, even when all the energy of a particle or photon is deposited in the detector the output-pulse amplitudes, from whatever component that constitutes the final stage, will show fluctuations around the maximum of the pulse-height distribution, that is assumed to represent the full radiation energy that is deposited within the detector. This is illustrated in Fig. 8-3 by the pulse-height spectra obtained for the γ rays from the decay of different radionuclides, using both NaI(Tl) and Ge(Li) detecting systems. The γ rays are monoenergetic to the highest degree, but the pulse-height spectra obtained by different detector systems show varying degrees of statistical spread of energy around the peak energy.

Most preamplifiers have statistically fluctuating electronic-noise levels equivalent to an output from a detector of an average of from 200 to 500 electrons. In other words, the average noise level introduced by the lower-noise preamplifier in a proportional-counter system would be equivalent to the deposition of radiation energy of about 7 keV in a pulse ionization chamber (for an average energy expended of 34 eV per ion pair), and of about 600 eV in a semiconductor detector. Further stages of amplification introduce relatively lower amounts of statistical noise.

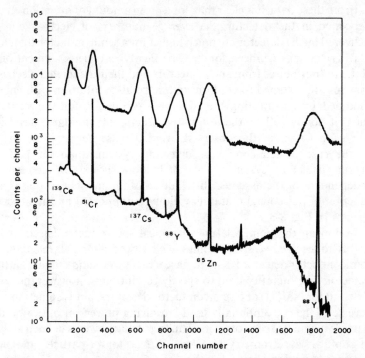

FIG. 8-3   γ-ray spectra of a mixed radionuclide source, consisting of 5 ml of solution in a glass ampoule. Upper spectrum obtained with the source within a 12.7-cm diameter NaI(Tl) well crystal. Lower spectrum obtained with the source at the face of a 60-cm³ Ge(Li) detector (from NCRP, 1978).

The initial interaction between an ionizing radiation and the detector is a statistical process in which given amounts of energy are expended in producing statistically varying effects in the detector. Thus, between 25 eV and 35 eV of energy are expended in creating an average of one electron-ion pair, for different types of radiation, in most gas ionization detectors; about 0.2 keV in a NaI(Tl) crystal per photoelectron from the cathode of the phototube; about 2 keV in liquid scintillators per photoelectron from the cathode of the phototube; and about 3 eV per electron-hole pair in a semiconductor.

From these data, it is clear that, for a given amount of radiation energy deposited in the detector, the average numbers of electron charges delivered by the detector are much higher for a semiconductor than for a NaI(Tl) crystal, or, indeed, for the two other types of detector mentioned. If the output pulses from such detectors and their associated electronic systems are assumed to conform to a normal distribution with energy, modified by substituting the observed average, $\bar{n}$, for the variance, then it follows that the estimated fractional standard deviation, $\bar{n}^{1/2}/\bar{n}$ or $1/\bar{n}^{1/2}$, of the observed average pulse height is much less in the case of semiconductor detecting systems than for NaI(Tl)-crystal detecting systems. This deduction should, in general, be independent of the assumed distribution of the number of pulses as a function of pulse height, and its validity is supported by the observed spectra of Fig. 8-3.

Following the nomenclature of optical spectrometry based on the ability to resolve two spectral lines of closely adjoining wavelength, λ, a semiconductor detector is said to have a better *resolution* than a NaI(Tl) detector. The quantity used to specify resolution is, however, not only different for NaI(Tl) and semiconductor detectors, but is sometimes the inverse of the conventionally used "resolving power" of optical and α-particle magnetic spectrometers. Thus resolving power is defined as $\lambda/\Delta\lambda$ in optical spectrometry and as $E/\Delta E$ for α-particle magnetic spectrometers, for α particles of energy E.

The resolution of NaI(Tl) detectors is usually given by $\Delta E/E$, expressed as a percentage, for the 662-keV γ ray in the decay of ^{137}Cs. In germanium γ-ray spectrometer systems, the resolution of the system is usually defined, however, as ΔE, the full width at half maximum (FWHM) of the photopeak distribution, at energy E. For Ge(Li) detectors larger than about $10\,cm^3$, the 122-keV γ ray in the decay of ^{57}Co and the 1.332-MeV γ ray in the decay of ^{60}Co are usually chosen for the reference energies. In the case of smaller semiconductor detectors, germanium or silicon used to detect lower-energy photons, the resolution ΔE is usually given in terms of lower-energy peaks such as the 5.9-keV and 122-keV γ rays from the decay of ^{55}Fe and ^{57}Co, respectively.

If we have a source of monoenergetic α particles, or γ rays, depositing their full energies within the sensitive volume of a detector, and the output pulses from the detector plus electronic system are recorded as a function of energy, then, instead of monochromatic lines, we obtain statistically broadened distributions as shown in the γ-ray spectra of Fig. 8-3. *If*, as has been generally the case, *it is assumed* that the shapes of these peaks, obtained with both the NaI(Tl) and Ge(Li) detecting systems, are described by a Poisson or normal distribution, then the resolution ΔE (FWHM) can be calculated from the normal distribution, given in Eq. 8-5.

The peak of the normal distribution, Eq. 8-5, is given by the value of P_x when $x = 0$ and is equal to $1/(2\pi\sigma^2)^{1/2}$. The value at half maximum is equal to P_x when $\exp(-x^2/2\sigma^2)$ is equal to one half, or when $-x^2/2\sigma^2 = \ln\frac{1}{2} = -0.69315$. Therefore, x is equal to $\pm 1.1774\sigma$, and the full width, $2x$, at half maximum (FWHM) is equal to 2.35σ. If we use the Poisson approximation $\sigma^2 = \bar{h}$, where $\bar{h}$ is the observed mean value of the pulse height h, corresponding to the photopeak, then the FWHM is equal to $2.35(\bar{h})^{1/2}$. But we have found that, for the detector systems considered, the pulse height is proportional to the number n of the charges collected in the detector, and that this, in turn, is proportional to the radiation energy E deposited within the sensitive volume of the detector. If we are considering the photopeak of a γ ray, then the constants of proportionality between h, n and E would be the reciprocals of their mean values, that correspond to the photopeak itself. Therefore, if Δh, Δn and ΔE are the full widths at half maximum of each variate, then $\Delta h/\bar{h} = \Delta n/\bar{n} = \Delta E/\bar{E}$. In practice the abscissa of a pulse-height spectrum may be in terms of h (i.e. MCA channel number), or of E where the detector plus MCA system has been calibrated, using various radionuclides emitting γ rays of well-known energy. Thus, if the distributions of h, n and E conform to normal distributions, the FWHM are given approximately by $\Delta h/\bar{h}$ equal to $2.35(\bar{h})^{1/2}/h$, or $2.35(\bar{h})^{-1/2}$, $\Delta n/\bar{n}$ equal to $2.35(\bar{n})^{-1/2}$, and $\Delta E/\bar{E}$ equal to $2.35(\bar{n})^{-1/2}$ or $2.35(\bar{h})^{-1/2}$.

In particular, for a NaI(Tl) detector and a photon with an energy E_γ, the FWHM of the full-energy peak, or photopeak, is given by

$$\Delta E_\gamma/E_\gamma = 2.35(\bar{n})^{-1/2}, \qquad (8\text{-}13)$$

whence

$$\Delta E_\gamma = 2.35(\bar{n})^{-1/2}E_\gamma, \qquad (8\text{-}14)$$

or

$$\Delta E_\gamma = 2.35(E_\gamma\bar{E}_p)^{1/2}, \qquad (8\text{-}15)$$

where $\bar{E}_p$ is the average energy required to produce one electron-ion pair or electron-hole pair, so that $\bar{n} = E_\gamma/\bar{E}_p$. The relationship given in Eq. 8-15 is that which would, if it were correct, be applicable to semiconductor-detector resolutions. It will be noted, however, from both Eqs. 8-13 and 8-14 respectively, that, for a given photon energy, the resolutions $\Delta E_\gamma/\bar{E}_\gamma$ for NaI(Tl) and ΔE_γ for semiconductor detectors are decreased (i.e. improved) for an increase in $\bar{n}$. It is also apparent from Eq. 8-15 that the smaller $\bar{E}_p$, the smaller is ΔE_γ.

It is necessary, however, to distinguish between the response of the detector and of the whole detector system. We have already noted that the resolution of the detector *system* is worse than that of the detector itself because of the statistical fluctuations introduced in the process of charge collection and by electronic noise in the processing of the output signal from the detector. The full width at half maximum of a normal distribution is $2.35\sigma^2$ or 2.35 times the variance. We have seen that the variances of independent distributions are added linearly (see the section "Propagation of Error" earlier in this Chapter). Therefore, dropping the factor 2.35 that is common to each term, the variance of the full width at half maximum of the photopeak can be expressed as a sum of variances, namely

$$\text{var (FWHM)} = \text{var } (n) + \text{var (charge collection)}$$

$$+ \text{var (electronics)}.$$

The variance of the charge collection is usually caused by recombination and electron trapping and, in the case of semiconductors, both of these effects have been markedly reduced in recent years by the production of better semiconductor material. Electronic performance has also been improved with reductions in noise levels by about a factor of ten since 1960 (compare, for example, the amplifier noise levels quoted in NCRP Reports 28 and 58, published in 1961 and 1978 respectively).

As we have mentioned, however, at the beginning of this section, the quantity of importance in promoting good resolution is the average number $\bar{n}$ of electron-ion or electron-hole pairs created, and collected, when a given amount of radiation energy is deposited within the sensitive volume of a detector. It is of interest therefore to determine the statistical fluctuations around the average value $\bar{n}$.

Experiments to investigate such fluctuations, carried out some thirty years ago, showed that the resolutions obtained with pulse ionization chambers were some two to three times better than values based, at that time, on the variance of the distribution.

The standard deviation, σ, given by Eq. 8-3 is a quantity that is not associated with any particular distribution, but in the case of the Poisson distribution (or of the normal distribution derived as an approximation of the Poisson), σ^2 is equal to the true mean or expectation value $E(n)$ of the distribution (see section B 3.4 and B 3.5 of NCRP Report 58). For a large number, k, of ionizing interactions in which an ionizing radiation produces a large number, n, of electron-ion pairs, we can use k for k $-$ 1 in the denominator of Eq. 8-4 and obtain an estimate of the variance that is equal to $\langle (n - \bar{n})^2 \rangle$; the angle brackets signifying the average of a moderately large number of measurements of n. We have also seen that for a Poisson distribution, or the normal distribution derived from the Poisson, an estimate of the variance is given by the observed average value of the distribution, and, therefore, the calculated estimate $\langle (n - \bar{n})^2 \rangle \sim \bar{n}$.

The discrepancy between the observed and expected resolutions of pulse ionization chambers was theoretically investigated by Ugo Fano in 1947, when he pointed out that the energy of a charged particle was transferred to gas molecules in a larger number of interactions than those that produced ionization. He stated that the "production of a measurable number of ionizations involves a large number of elementary processes" or interactions in which the charged particle not only ionized the gas molecules, but also *excited them without ionization*. He then derived a factor F, now known as the Fano factor, such that

$$\langle (n - \bar{n})^2 \rangle = F\bar{n} \tag{8-16}$$

where $\bar{n}$ was equal to the energy of the ionizing particle divided by the experimentally measured average energy expended in producing an electron-ion pair. The experimentally determined variance on the left-hand side of Eq. 8-16 is not dependent upon the assumption of any distribution, but if it is found that $F = 1$, then the statistical fluctuations must conform to the Poisson distribution. Empirical values for F indicated, however, that the variances of the distribution of n for pulse ionization chambers were about 0.3 to 0.5 times those given by the Poisson distribution.

That the fluctuations in the number of electron-ion pairs do not conform to a Poisson distribution is not wholly surprising, because the probability that, say, a full-energy photoelectron interacting with a gas molecule will produce primary and secondary electron-ion pairs is not small. If the binomial distribution were able to fit the distribution better, this would give the same mean number of ion pairs corresponding to the full-energy photoelectron, as would the Poisson distribution, but the variance of the binomial distribution would be equal to the variance of the Poisson distribution multiplied by $(1 - p)$ which would correspond to the probability of an interaction, or collision, that results only in an excitation and not ionization. If the probability p of producing ionization becomes much less than one, then the variances of the binomial and Poisson distributions become equal. As the factor $(1 - p)$ is clearly less than one it could correspond to the Fano factor. The situation is, however, even more complicated in that correlations exist between the outcomes of successive interactions of a photoelectron, both in gas ionization chambers and semiconductor detectors. Such interactions arise because a series of exciting or ionizing interactions can reduce the energies of primary charged particles and secondary, cascade, electrons below the ionization threshold potential, so that the residual energies can be dissipated only by excitation. It appears that such interactions may further reduce the Fano factor, that, at this time is, however, probably most often determined by experiment.

In the case of semiconductor detectors, the Fano factor is introduced into Eq. 8-15 to give

$$\Delta E_\gamma = 2.35(FE_\gamma \bar{E}_p)^{1/2} \tag{8-17}$$

Fano factors as low as 0.06 and 0.07 for Ge(Li) and Si(Li) detectors, respectively, have been measured by H. R. Zulliger and D. W. Aitken. These authors also make the interesting observation that for both germanium and silicon, the Fano factor has decreased from about 0.4 or 0.5 in 1964 to about 0.06 or 0.07 in 1969. Their article also gives references on the subject for further reading. The Fano factor has also been discussed by W. van Roosbroeck (1965).

With such low values being measured for the Fano factor, it is clearly inappropriate to use the normal-distribution function to fit the peaks obtained with semiconductor γ-ray spectrometers. For this purpose, a Gaussian function in the form $Ae^{-(x-\bar{n})^2/B}$, where A is the peak maximum, and B is related to the peak width, is generally used in the peak-fitting of spectral peaks by computer (see NCRP Report 58). The resolution is still obtained from the full width at half maximum of the experimental peak.

Two examples of the use of the Fano factor, F, may help to explain its practical application.

As our first example we can consider an ionization chamber containing argon and operating in the pulse mode. The value of W for argon is 26.5 eV, so that an ionizing radiation of, say, 500 keV depositing all its energy in the gas of the chamber will give rise to about 18,900 electron-ion pairs. The resolution $\Delta E/E$, given by $2.35\bar{n}^{-1/2}$ is equal, therefore, to 0.0172 or 1.72 per cent. But the experimentally determined value of F for argon is approximately 0.2, which gives a resolution of 2.35 $(0.2/\bar{n})^{1/2}$, i.e. 2.35 $(F\bar{n})^{1/2}/\bar{n}$, equal to 0.0077 or 0.77 per cent. This represents an improvement in resolution, $\Delta E/E$, by more than a factor of 2. As E is equal to 500 keV, ΔE, the full width at half maximum of the peak obtained as a result of many 500 keV events depositing all their energy in the detector, would be equal to 3.85 keV.

The second example can be that of a semiconductor detector, in which 500-keV photons are being completely absorbed in the sensitive volume. We will assume that for such a detector $\bar{E}_p$, the average energy expended in producing an electron-hole pair, is equal to 2.96 eV. Then the average number of such pairs produced by one 500-keV event would be approximately equal to 169,000. Using an experimentally determined Fano factor of 0.13, the energy resolution given by Eq. 8-17 is approximately 1 keV FWHM, or 0.21 per cent of E_γ. This is nearly four times better than the resolution of the argon-filled pulse ionization chamber, because of the much larger number, $\bar{n}$, of charge carriers produced.

For detectors in which more than one process is involved in developing the output signal, the resolution is, as mentioned earlier, a function of more than one statistical distribution. Examples of such detectors are gas proportional counters and scintillation detectors. In the former the production of ion pairs is followed by gas multiplication, and in the latter the light output signal from the scintillator undergoes conversion to electrons at the photocathode of the phototube, with subsequent amplification of these electrons at each dynode of the phototube. All these processes increase the total statistical variance of the output signal. In the case of a typical gas proportional counter, where just the processes of ionization and multiplication are involved, the resolution is decreased typically by about a factor of 2, the FWHM at 500 keV being about 8 keV as compared with the 3.8 keV for the argon-filled pulse ionization chamber, discussed earlier. The calculation of the resolution for scintillation detectors is much more complex because of the many intermediate processes involved. The subject has, however, been reviewed by Birks (1964) and Adams and Dams (1970).

From this brief discussion, it has clearly emerged that of the detectors in general use, the semiconductors give the best resolution and the scintillator systems the poorest. This is dramatically illustrated by again referring to Fig. 8-3 where the γ-ray spectrum of a mixed-radionuclide source obtained with a NaI(Tl) detector is compared with that obtained for the same source with a Ge(Li) detector.

Dead-Time Corrections

As mentioned in Chapter 7, there exists for every pulse measurement system a finite dead time. This dead time is imposed every time a pulse is processed by the system, requiring a correction to be made for those pulses, or counts, that are lost as a result. Prior to calculating any corrections to the observed count rate for dead-time losses it is, of course, first necessary to determine both the value of the dead-time interval, τ, and whether it is extending or nonextending (see Dead-Time Losses and Circuits in Chapter 7).

Considering a system with a nonextending dead time, if the *observed* count rate, in counts per second, is n, then the time per second that the system is insensitive (or "dead") is $n\tau$, when τ is also in seconds. It follows that the system is receptive to counts (or "live") for a fraction $1 - n\tau$ of the counting interval. Therefore, the dead-time corrected count rate, N, is given by

$$N = n/(1 - n\tau), \qquad (8\text{-}18)$$

for a system with a nonextendable dead time and an input rate that does not vary significantly during the measurement.

For a system with an extending dead time, τ, it is necessary to know the statistical distribution of the time intervals between successive events. Because radioactive decay is a random process, the input events arrive in a random time sequence. As we have stated previously the time interval distribution of these events can be described by the Poisson distribution. From this the probability that no events occur within the dead-time interval, τ, can be shown to be $e^{-N\tau}$, where N is the mean input rate, and the observed count rate, n, is

$$n = Ne^{-N\tau}. \tag{8-19}$$

From this equation the input event rate (or corrected count rate), N, can not be obtained in terms of the observables, namely, n and τ. The following approximation for Eq. 8-19 can, however, be used

$$N \approx n/(1 - n - (n\tau)^2/2) \tag{8-20}$$

to obtain corrected count rates with accuracies of about 0.1 per cent for extendable dead-time losses of 10 per cent or less. For corrections of higher accuracy, it is necessary to use iterative computational procedures (see Section 2.7 of NCRP Report 58).

The relationships between the observed count rates, as a function of the input event rate, are shown in Fig. 8-4 for both types of system dead times. The observed and input rates are plotted in units of τ^{-1} (i.e., $n\tau$ vs. $N\tau$). For example, for a dead time, τ, of 1 μs, a value of N of 10^6 s^{-1} would correspond to 1 on the $N\tau$ axis and the corresponding value of n, for a nonextendable dead time would be 5×10^5 s^{-1}.

As illustrated in Fig. 8-4, for a system with a nonextendable dead time, the observed count rate approaches asymptotically to τ^{-1} (i.e. $n\tau = 1$) for input event rates considerably in excess of τ^{-1}, but the system is never rendered inoperative or "paralyzed" by excessive input-event rates. On the other hand, as shown in Fig. 8-4, systems with an extendable dead time reach a maximum observed count rate for an input rate of τ^{-1} (i.e. $N\tau = 1$), which decreases to zero for input rates greatly in excess of τ^{-1}. As, for each value of observed count rate, n, there exist two possible values of input event rate, except for the single value of n corresponding to an input event rate of τ^{-1}, the possibility of misinterpreting the input

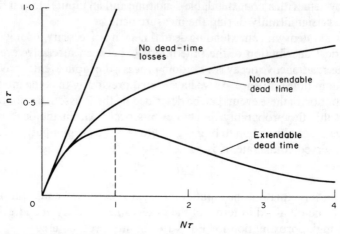

FIG. 8-4   Relationship between observed count rates, n, and input-event rate, N, for extendable and nonextendable dead times (after NCRP, 1978).

count rate by using observed count rate data from unexpectedly strong sources cannot be ignored. One way to check for this potential source of error is to decrease the input-event rate by some means, such as increasing the source-to-detector distance and comparing the computed values of N obtained using the two observed count rates.

Because of the inexact relationship for calculating the corrected count rates and the possible ambiguity between the observed count rate and the "true" input rate, systems in which an extending dead time dominates are to be avoided. It is for these reasons that systems with a dominating nonextendable dead time are recommended in Chapter 7.

Regardless of the type of system dead time, the calculated value of N is dependent on the *observed total* count rate. It is extremely important, therefore, that the correction for dead-time losses be made *before any* other correction to the observed data be made, such as, for example, subtraction of the background count rate.

Background Correction

Any radiation detection system will give a response, even in the absence of a specific radioactive sample. This background response of the detector is due to cosmic radiation and to radiation emanating from naturally-occurring radioactivity that is always present in construction materials, such as concrete, brick, aluminum, lead, copper and steel. Although such background contributions to the observed detector response are always present, they may be minimized through the use of shielding around the detector and by careful selection of the construction materials used for the building, such as low-activity aggregate in the concrete, the shielding and detector itself.

The background count-rate response of a radiation-detector system, r_b, is usually measured with an "inactive" blank substituted for the radioactive source. The blank source is used to duplicate the effect of the radioactive source in attenuating and scattering the background radiations. For the best measure of the background correction, the blank source, and measuring conditions should be as identical as possible to those for the radioactive-source measurement. Also, the background determination should be made either immediately preeceding, and/or following the radioactive source measurement.

By making the background measurement close in time to the source measurement, errors due to short-term fluctuations in the background can be avoided. By making background measurements at longer intervals, long-term variations in the performance of the measuring system, such as increased noise contributions or detector contamination, may be uncovered.

Once the background response per unit time, r_b, has been determined it is subtracted from the source plus background response per unit time, r_{s+b}, of the detector system, be it an ionization chamber or a pulse-rate detector. In the case of a pulse-rate detector the background contribution must be subtracted, as mentioned in the last section, *after* the count rate has been corrected for dead-time losses.

Source-Detector Geometry

Radiations from a radioactive source are emitted isotropically. In the absence of scattering, the amount of this radiation incident upon the "end window" of a cylindrical detector is proportional to the solid angle Ω, in steradians, which the detector window subtends at a point source. If the source is on the axis of the cylindrical detector, as shown in Fig. 8-5, the solid angle is given by

$$\Omega = A_s/R^2, \tag{8-21}$$

where the area, A_s, of the sphere of radius R, that intersects the periphery of the end-window of the detector, is given by

$$A_s = 2\pi R^2 (1 - \cos \theta), \tag{8-22}$$

θ being the half angle of the cone subtended by the detector window to the point source, as shown in Fig. 8-5. As the area of the sphere is $4\pi R^2$,

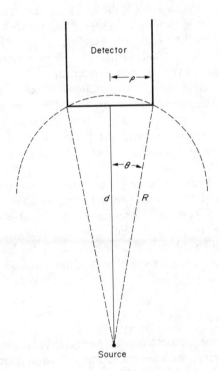

FIG. 8-5 "Geometry" for a point source relative to a right cylindrical end-window detector "face-on" to the source.

the fraction of the radiation incident upon the detector window is

$$G = A_s/4\pi R^2, \qquad (8\text{-}23)$$

or
$$G = \tfrac{1}{2}(1 - \cos\theta), \qquad (8\text{-}24)$$

where G is the *source-to-detector geometry*. G is also variously referred to as the *geometry factor, detector geometry, geometrical efficiency*, and *extrinsic detector efficiency*. When θ is equal to $\pi/2$, Eqs. 8-21 and 8-22 give Ω equal to 2π steradians and for θ equal to π, Ω is equal to 4π steradians.

The source-to-detector geometry, as a function of solid angle, is given by Eqs. 8-21 and 8-23 to be

$$G = \Omega/4\pi. \qquad (8\text{-}25)$$

When θ is equal to π, the sensitive volume of the detector completely surrounds the source, the geometrical efficiency is unity and the solid angle subtended by the detector window to the source is equal to 4π steradians. Such detectors are, therefore, generally referred to as "4π" detectors. Likewise many other detectors are characterized by the solid angle that they subtend to the source, and often, also, by the type of radiation that they are routinely used to detect, such as $0.8\pi\alpha$, $2\pi\beta$, and so on.

If d is the distance between the point source and the detector window, and ρ is the radius of the window, then, for θ small but finite and increasing values of d, the area A_s tends to equality to $\pi\rho^2$ and d itself tends to equality to R. Equation 8-23 can then be replaced by the approximation

$$G = \rho^2/4d^2. \qquad (8\text{-}26)$$

This is the well-known *inverse-square law*. We have derived this law for a point source, but even when the source is small and finite in size, provided that θ is small and d is reasonably large, the inverse-square law represents a good approximation for the variation of geometrical efficiency with source-to-detector distance.

More complex relationships that apply to other geometries, such as disc or cylindrical sources, or point sources off the axis of a cylindrical detector, have been discussed, for example, by A. H. Snell.

Radioactive-Decay Calculations

During the time intervals in which counting data are being collected, the activity of a radioactive source is continuously decreasing. Therefore, the count rate for any but long-lived radionuclides cannot be calculated simply by dividing the total number of accumulated counts by the duration of the counting interval. Activity, and hence count rate, are functions of time and therefore only have practical significance at a given time, t, and for a given decay constant, λ, when the activity or count rate at a previous time, t, is known (see Chapter 2). What then is the correct reference time that should be assigned to a count rate that has been measured during an interval of time T? When T is short compared with the half life of the radioactive source, it is generally immaterial what reference time be chosen within the counting interval. If T is, however, not short compared with the half life, then, as we shall see, the reference time can be taken, with very good approximation, as the midpoint of the counting interval.

If r_0 is the mean count rate at the beginning of a counting interval, of duration T, and r is the count rate derived from the total number of observed counts in the interval T divided by T, then the total number of counts observed in time T is

$$rT = r_0 \int_0^T e^{-\lambda t}\, dt,$$

$$= r_0(1 - e^{-\lambda T})/\lambda,$$

and, therefore,

$$r_0 = r\lambda T/(1 - e^{-\lambda T}). \tag{8-27}$$

Thus r_0 for a reference time at the beginning of the counting interval can be readily calculated, provided λ is known. As, however, such decay calculations are only needed for relatively short-lived radionuclides, λ can normally be derived from a half-life measurement carried out over successive counting intervals, taking the reference time at any, but the same, part of the counting interval (often at the beginning of the interval). (Decay calculations are often called "corrections".)

Mean counting rates are also often assigned a reference time at the midpoint of a counting interval. Such a procedure introduces very little error, however, even when, as we will show, the interval T is as large as 10 per cent of the half life. The observed mean count rate, r, derived from the total counts accumulated in the counting interval T is related to r_0, the mean count rate at the beginning of the counting interval, by the equation, from Eq. 8-27,

$$r = r_0(1 - e^{-\lambda T})/\lambda T. \tag{8-28}$$

At the midpoint of the counting interval, the mean count rate r_m is given by

$$r_m = r_0 e^{-\lambda T/2}. \tag{8-29}$$

The ratio of the two mean count rates is, therefore,

$$r/r_m = (1 - e^{-\lambda T})/(\lambda T e^{-\lambda T/2}). \tag{8-30}$$

If $T_{1/2}$ is the half life of the radionuclide being measured, then λ is equal to $0.69315/T_{1/2}$. When T is as large as one half of one half life, then T is equal to $0.69315/2\lambda$ which, on substitution in Eq. 8-30, gives r/r_m equal to 1.005. In other words the "true" mean count rate at the mid-time of the counting interval is different from the observed mean count rate, r, by only one-half of one per cent. A similar substitution into Eq. 8-30 of the counting interval T equal to one-tenth of a half life, shows that r differs from r_m by only 0.02 per cent.

Before making decay calculations, the observed count-rate data must first be corrected for dead-time losses and the background count rate, in that order.

An expression that combines the corrections for dead-time losses, background and radioactive decay is quoted in Section 2.7.3 of NCRP Report 58.

Direct Measurements of Radioactivity

The difference between direct and indirect measurements of radioactivity is that direct measurements are made entirely in terms of physical quantities such as length and time, whereas indirect measurements are made in terms of either (i) another calibrated radioactivity source (i.e. a *radioactivity standard*), of the same radionuclide, or (ii) an instrument that has been pre-calibrated by means of a standard of that radionuclide or by means of a number of standards of different radionuclides emitting radiations of different energies.

One of the oldest methods of standardization is that using a detector with a defined solid angle. This is essentially based on application of the geometry shown in Fig. 8-5. The solid angle subtended by the detector at the source is calculable from the source-to-window distance d and the radius of the window ρ. Then, provided that the radiation from the source suffers no scattering or other attenuation and that every radiation entering the detector is recorded, the activity of the source is simply determined by the number of recorded counts per second divided by the geometry G, as given by Eq. 8-25. Practical realization of the geometry illustrated in Fig. 8-5 is often in the form of a radioactive source and a detector behind a circular diaphragm, the whole structure being rigidly mounted in a vacuum vessel. The dimensions ρ and d then refer, respectively, to the radius of the diaphragm aperture and the orthogonal distance from the point source to the plane of the surface of the diaphragm. Thus the source might be an α-particle-emitting nuclide, deposited very thinly from a drop of solution upon a thin source mount in order to minimize source self-absorption and scattering, with either a thin layer of ZnS(Ag) spread on a plastic disc or a thin CsI(Tl) crystal viewed by a phototube comprising the detector. Alternatively, for example, a thin β-particle source might be mounted in the vacuum vessel, with a thin-window gas proportional counter behind the diaphragm.

For such α- and β-particle measurements, the solid angle subtended by the diaphragm aperture at the source is usually less than 1π and may be as small as 0.1π. Walter Bambynek has used defined-solid-angle counters for the metrology of photon-emitting nuclides with photon energies in the range of 1 to 80 keV. He states that, for the successful application of this method, the radiations should be "lightly scattered but heavily absorbed."

In contrast to these low solid-angle detectors there are the 4π-steradian detectors illustrated in Figs. 6-9, 6-10 and 6-11. These are assumed to have intrinsic efficiencies of 100 per cent (i.e. $\varepsilon = 1$) for thin, or so-called "weightless," sources of energetic α and β particles. In practice the efficiencies of such detectors have indeed been found to be very close to 100 per cent for β-emitting nuclides such as ^{32}P and

^{90}Sr–^{90}Y. Another variation of the 4π detector is the internal gas proportional counter, used for the standardization of such radioactive gases as ^{3}H, ^{14}CO$_2$, ^{37}Ar, ^{85}Kr and ^{133}Xe, the gas in question being mixed quantitatively with the counting gas, and counted inside the counter. Such counters are usually of cylindrical geometry and corrections must be made for radiations lost to the walls, and for the drop in electric field at each end of the counter where the anode wire passes out of the counter through an insulator. The wall correction is usually made by counting at different gas pressures and then plotting count rate against reciprocal pressure and extrapolating to zero reciprocal pressure, i.e. to infinite pressure. The end losses are usually eliminated by counting in two counters of different lengths but with otherwise identical dimensions, at the same gas pressure. The difference in the two counting rates gives the count rate for an "ideal" counter having an effective length equal to the difference in the lengths of the two counters, and with a uniform electric field.

Liquid-scintillation counters are also essentially 4π detectors, but with efficiencies rarely greater than 90 per cent. They can, however, be used very successfully as high-efficiency detectors both for comparative measurements of radioactivity, and, together with a γ-ray detector, for direct measurements using the method of coincidence counting to be described in this section.

The first radioactivity standards were those of radium prepared by Marie Curie and Otto Hönigschmid in 1911 by sealing carefully weighed quantities of radium chloride into thin-walled glass tubes. These were "direct" mass standards, but, as mentioned in Chapter 2, the mass of ^{226}Ra is related to its activity through its decay constant. Thus using a half life of 1600 years we derived an activity of $3.66 \times 10^{10}\,\text{s}^{-1}$ per gram of ^{226}Ra.

The radioactive decay law is also the basis for the measurement of radioactivity by mass spectrometry, by means of which the activity is determined by measuring the number of radioactive atoms comprising a given quantity of the source and then measuring the half life of the source. The activity is then calculated from the decay law, $dN/dt = -\lambda N$. Conversely, the half life of a long-lived radionuclide can be measured by determining the activity of a source of the radionuclide, of known mass and isotopic abundance. In this manner, the half life of ^{14}C was found to

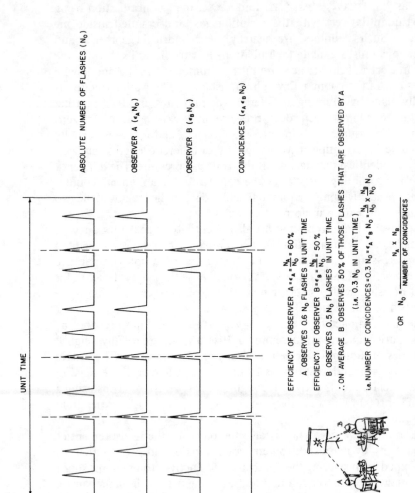

FIG. 8-6   Elementary example illustrating the coincidence-measurement technique, as applied to *two observers*.

be 5730 years. Half lives have also been measured mass-spectrometrically by determining the growth in the number of daughter atoms, for a given number of parent atoms initially present. Thus the half lives of ^{3}H and ^{239}Pu have been determined by measuring, respectively, the growth of ^{3}He and ^{235}U.

Probably the most powerful method of measuring the activity of a radioactive source is by the method of coincidence counting. This is said to have been first used by Rutherford to measure the efficiencies of two observers looking through low-power microscopes at the same ZnS(Ag) screen upon which α particles from one radioactive source were incident. Each time that an observer saw a flash of light, he pressed a switch that caused a record to be made upon a moving strip of smoked paper. This is illustrated in Fig. 8-6, where the observant reader will note that the observer with the stein is slightly less efficient than he with the coffee! It is also to be noted, however, that in this application only the efficiencies of the observers are measured and not the intrinsic efficiency of the detector relative to the source.

However, in like manner, if we have a source of a radionuclide that is emitting two different types of radiation, that are independently counted by two different detector systems, with a coincidence circuit to record events that are coincident in time in the two detectors, as shown in Fig. 8-7, then the coincidence equations can be used, or eliminate,

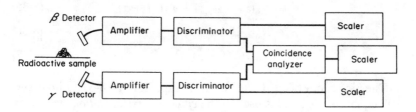

FIG. 8-7   Block diagram of a basic β–γ coincidence counting system.

the intrinsic efficiencies of the two detectors to the respective radiations. Thus, if we had a source of ^{60}Co, that emits β and γ rays in coincidence, and the former are counted by a β-particle-detector system with an efficiency ε_β, and the latter are counted by a γ-ray-detector system with an efficiency ε_γ, then the equations that correspond to the coincidence equations in Fig. 8-6 are

$$r_\beta = N_0\varepsilon_\beta, \tag{8-31}$$

$$r_\gamma = N_0\varepsilon_\gamma, \tag{8-32}$$

and $$r_c = N_0\varepsilon_\beta\varepsilon_\gamma, \tag{8-33}$$

whence $$N_0 = r_\beta r_\gamma/r_c. \tag{8-34}$$

Equation 8-34 gives the simple basic coincidence equation, but, in practice, corrections must be made for the sensitivity of the β-particle detector to γ radiation and conversion electrons, for random coincidences, for dead-time losses, for background, and, if necessary, for decay. The method of coincidence counting was first used to measure the activity of a radioactive source by H. Geiger and A. Werner in 1924.

As a β detector is sensitive to conversion electrons and as its efficiency ε_β is normally less than one, it is necessary to modify the coincidence equations for radionuclides that decay with the emission of conversion electrons. In such cases these electrons can be detected by the β detector during the fraction, $(1 - \varepsilon_\beta)$, of the decays that the β particles are not detected, thus increasing the β-channel count rate. If α is the internal-conversion coefficient and ε_{ec} is the conversion-electron efficiency of the β detector, then the relevant coincidence equations are

$$r_\beta = N_0(\varepsilon_\beta + \alpha\varepsilon_{ce}(1 - \varepsilon_\beta)/(1 + \alpha)), \tag{8-35}$$

$$r_\gamma = N_0\varepsilon_\gamma/(1 + \alpha), \tag{8-36}$$

$$r_o = N_0\varepsilon_\beta\varepsilon_\gamma/(1 + \alpha) \tag{8-37}$$

The β efficiency of the β detector (from Eqs. 8-36 and 8-37) is, however, still equal to r_c/r_γ, and is used in the *efficiency extrapolation method*, in which the observed value of $r_\beta r_\gamma/r_c$ is plotted against a suitable function, f, of $\varepsilon_\beta(= r_c/r_\gamma)$, the value of ε_β being varied, either by redissolving the source on the source mount using a drop of water, by placing thin films sequentially over both sides of the source, or by changing the level of discrimination in the β channel. The forms taken for the function f are usually $f' = \varepsilon_\beta$, or $f'' = (1 - \varepsilon_\beta)/\varepsilon_\beta$. The extrapolated value of $r_\beta r_\gamma/r_c$ at either $f' = 1$ or $f'' = 0$, according to which abscissa has been chosen, is then equal to N_0, because the β detector at $\varepsilon_\beta = 1$ (i.e. for $f' = 1$, or $f'' = 0$), would be detecting all β events, and conversion electrons and γ rays depositing energy in the β detector would now be in coincidence with true β counts.

More detailed descriptions of this method and of the simple coincidence method itself, including the methods of correlation and anti-coincidence counting, together with details of the various corrections that must be made, will be found in NCRP Report 58 and in the Proceedings of the First International Summer School on Radionuclide Metrology held at Herceg Novi. Their further discussion is, however, somewhat beyond the scope of this book.

The last direct method of standardization to be discussed is that of sum-peak coincidence counting using a single NaI(Tl)-spectrometer system. This method can be used only for radionuclides decaying with two or more photons in coincidence and with no direct transitions from the radioactive parent to the ground state of the daughter nucleus.

It is necessary first, however, to understand what processes are involved in producing a γ-ray spectrum such as that shown in Fig. 8-3. This is best done by considering a radionuclide source emitting monoenergetic γ rays of, say, 500 keV that interact with a NaI(Tl)-spectrometer system comprising a NaI(Tl)-crystal detector, amplifier, and multichannel analyzer as described in Chapter 7 on Electronic Instrumentation. In such a system the multichannel analyzer sorts and records the pulse height for each ionizing event occurring in the detector, the pulse height being directly proportional to the energy deposited in the detector by that ionizing event.

If we had an *ideal* NaI(Tl)-detector system, all of the γ rays would be totally absorbed and we would obtain output pulses from the last dynode of the phototube that were all of equal amplitude, corresponding to 500 keV, as shown in Fig. 8-8a. This presupposes that every γ ray produces exactly the same number of photons of visible light for every detected γ ray, and that these photons are all transmitted to the photocathode of the phototube where each photon produces equal numbers of photoelectrons, and that each dynode of the phototube produces the same yield of secondary electrons per electron incident upon it. In other words the resolution is perfect and $\Delta E/E$ is equal to zero.

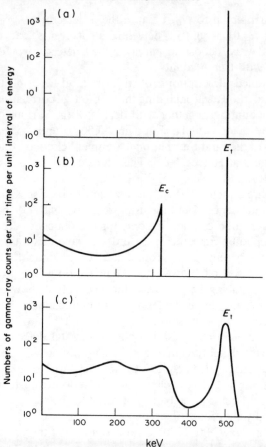

FIG. 8-8 γ-ray spectrum of a source emitting monoenergetic 500-keV γ rays: (a) obtained with an *ideal* NaI(Tl) detector system; (b) including the Compton interactions with the detector; (c) a typical observed spectrum.

This does not, in fact, happen as we are well aware, there being statistical fluctuations at every "interface" from high-energy photons to low-energy photons (of light), low-energy photons to photoelectrons, and photoelectrons to secondary electrons at each dynode.

Let us now consider, however, the kind of output from the NaI(Tl)-detector system if 500-keV γ rays had either single Compton, or photoelectric interactions within the crystal. In a single photoelectric interaction within the crystal, a γ ray would give up all its energy (less a small amount of atomic-electron binding energy), and the photoelectron would have an energy of essentially 500 keV. The photoelectrons would be raised to the conduction band and would travel through the crystal stimulating lattice vibrations (phonons) and the emission of light. If the light yield were the same for every 500-keV photoelectron, and, if there were no statistical fluctuations from this point on to the output from the last dynode of the phototube, we would again obtain output pulses of all the same amplitude as indicated by the single ordinate at 500 keV in Fig. 8-8b. Again $\Delta E/E$ is zero.

But what happens to the single Compton interactions? These, as we shall see, cannot deposit all of their 500 keV of energy within the crystal in a single interaction.

From Eq. 3-30, the energies of the scattered photons are given by

$$hv' = hv/(1 + hv(1 - \cos\theta)/m_0 c^2).\qquad(8\text{-}38)$$

Also, if the γ ray is scattered through 180° thereby transferring a maximum amount of energy to the electron

$$hv' = hv/(1 + 2hv/m_0 c^2).\qquad(8\text{-}39)$$

The rest mass of the electron, $m_0 c^2$, is equal to 511 keV, so that the maximum amount of energy, say, that a 2.5-keV photon could transfer to an electron in a single collision would be about 1 per cent of its initial energy, hv. On the other hand when hv becomes very large compared with $m_0 c^2$, we can neglect 1 by comparison with $2hv/m_0 c^2$ in Eq. 8-39, and hv', the energy of the back-scattered photon, approaches equality to $m_0 c^2/2$, or 256 keV. A 125-keV γ ray back-scattered through 180° is left with 90 keV of energy. Thus we obtain a very useful generalization, to which we will later return, that all back-scattered γ rays with energies above about 125 keV, have energies in the range of about 100 to 250 keV after scattering.

Whenever a photon is scattered through 180° and suffers a maximum loss of energy and then escapes, the recoil electron acquires this energy, E_c. Photons scattered through less than 180° will impart less energy to the recoil electron, and thence to the crystal. Thus there will be a continuum of energies, from 0 to E_c, imparted to the crystal, as shown in

Fig. 8-8b. Because of the variation of scattering cross-section of γ rays by electrons, the shape of the electron spectrum between 0 and E_c is approximately as depicted in Fig. 8-8b for the single scattering of 500-keV photons.

If we now return completely to real life the idealized "theoretical" spectrum turns into the experimental spectrum of Fig. 8-8c. Here the total energy peak has been statistically broadened by each of the processes involved in the train of events from the γ-ray interaction with the crystal to the final output pulse from the phototube. There may also be one or more Compton scatterings before the γ ray escapes from the crystal so that the sum of the energies of several recoil electrons may give a total energy between E_c and E_t. (The time interval between consecutive interactions of a γ ray moving with the velocity of light are so short that the energies of the recoil electrons are summed in one output pulse.) On the other hand the γ ray may not escape from the crystal, but after giving up energy in one or more Compton interactions may impart the remainder of its energy to a photoelectron. Thus this γ ray will be recorded at pulse height E_t, corresponding to the total energy peak, which is usually referred to as the *photopeak* or *full-energy peak*. The relative numbers of events recorded in the photopeak to those in the "Compton continuum" will depend on the size of the crystal and E_t. The relative number of events under the photopeak to the total number of events in the whole spectrum is usually called the *photopeak-to-total ratio*. The photopeak efficiency is often referred to as ε_p and the total efficiency as ε_t. If a γ ray undergoes more than one or two scattering interactions, its energy becomes so reduced that it will more probably surrender all its remaining energy in a photoelectric interaction with an atom rather than escaping from the crystal. By reference to Figs. 3-15 and 3-16 it is seen that attenuation by photoelectric effect is more probable than by the Compton effect at low energies. The curves shown in these figures are for low- and high-Z materials, aluminum and lead. The attenuation coefficients for sodium iodide will be intermediate between those of aluminum and lead (see Fig. 41 in NCRP Report 58).

The "back-scatter peak" shown in Fig. 8-8c at about 170 keV is an artifact that may or may not be present depending on the proximity of scattering material, such as, for example, background shielding material around the detector and a source. As mentioned earlier, γ rays from the source will be back-scattered by such material with an energy of about 100 to 250 keV, and they can then deposit all their remaining energy in the NaI(Tl) crystal.

The spectrum shown in Fig. 8-8c is typical of that which would be obtained with a monoenergetic γ-ray-emitting nuclide, using a NaI(Tl)-spectrometer system. If, however, the spectrometer system is used to record the spectrum of the ^{94}Mo γ rays emitted in the decay of ^{94}Nb we would normally obtain a spectrum which would be similar to that shown in Fig. 8-9 for γ-ray energies below 1.2 MeV. This spectrum would be typical for a low-geometry detector, such as, say, a 2 × 2-inch or 3 × 3-inch crystal. The probability of both γ rays being detected

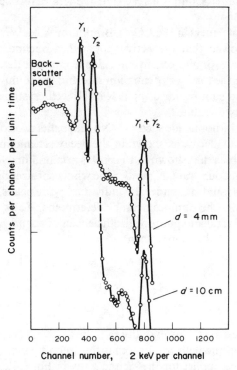

FIG. 8-9   Sum-peak spectra obtained with two 3″ × 3″ NaI(Tl) crystals separated by respectively 4 mm and 10 cm. The energies of the γ_1, γ_2 and $\gamma_1 + \gamma_2$ peaks are 702.63, 871.10 and 1573.73 keV (after R. L. Heath).

simultaneously in the crystal is proportional to the product of the probabilities of each γ ray being detected. If the geometry is increased to nearly 4π by sandwiching the source between two detectors, or by using a NaI(Tl) well crystal, a third peak appears in the spectrum at 1.57 MeV, which represents the summation of the energy of two coincident γ rays that are totally absorbed in the crystal. In Fig. 8-9 two sum peaks are shown, the larger peak for a ^{94}Nb point source located 4 mm away from the front face of each of two 3 × 3-inch NaI(Tl) crystals, and the smaller peak for the source-to-detector distance increased to 10 cm. It will be noted that not only do the two totally absorbed individual coincident γ rays sum to form the sum peak, but that the Compton spectra also sum to form a summed Compton spectrum at energies just below that of the sum peak.

In a series of papers in 1963, G. A. Brinkman, A. H. W. Aten, and J. Th. Veenboer showed that the activity of such a nuclide, that emitted a number of γ rays, or photons, in coincidence, could be determined simply by measuring the total γ-ray emission rate and the count rates under the individual and sum peaks, using a NaI(Tl) well crystal, of as nearly 4π "geometry" as possible.

Consider a radionuclide such as ^{94}Nb that emits two coincident γ rays, γ_1 and γ_2, in its decay, as shown in the decay scheme in Fig. 8-10. If a spectrum such as that shown in Fig. 8-9 is accumulated for a given time then, the total count rate, N_t, under the whole spectral distribution, the count rates, A_1 and A_2, under the individual γ-ray peaks, and the count rate, A_{12}, under the sum peak can be determined. If ε_t', ε_t'', ε_p', and ε_p'' are the respective total and photopeak efficiencies of γ_1 and γ_2, and N_0 is the activity of the ^{94}Nb source, then

$$A_1 = N_0\varepsilon_p'(1 - \varepsilon_t''), \tag{8-40}$$

$$A_2 = N_0\varepsilon_p''(1 - \varepsilon_t'), \tag{8-41}$$

and $$A_{12} = N_0\varepsilon_p'\varepsilon_p''. \tag{8-42}$$

Eq. 8-40 represents the probability that the first γ ray, γ_1, will deposit *all* of its energy in the detector, and that the second γ ray, γ_2, will deposit *none*. The same applies for the second γ ray in Eq. 8-41. Equation 8-42 represents the ordinary condition for a coincidence in which all of the energy of both γ rays will be deposited in the detector.

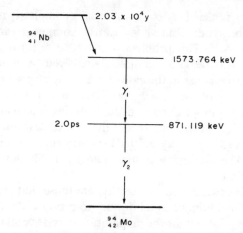

FIG. 8-10 ^{94}Nb decay scheme. For a description of the methods used to evaluate γ-ray energies and for further references, see R. G. Helmer, R. C. Greenwood and R. J. Gehrke (1978).

A further possibility is that neither γ ray interacts with the crystal, and will therefore be unrecorded in the pulse-height spectrum. This is represented by

$$N_0 - T = N_0(1 - \varepsilon_t')(1 - \varepsilon_t''), \qquad (8\text{-}43)$$

where T is the total count rate, obtained by counting all output pulses from the detector in a given time at each of several discriminator bias settings and extrapolating to zero bias voltage. By the nature of the spectrum, the ordinates of which are counts per unit increment in energy, the values of A_1, A_2, and A_{12} are equal to the areas under their respective peaks, above the Compton and background continuum. No dead-time-loss corrections need be made if the multichannel analyzer is operated in the live-time mode.

By substituting for $1 - \varepsilon_t'$ and $1 - \varepsilon_t''$ from Eq. 8-40 and 8-41 into Eq. 8-43, and eliminating $\varepsilon_p'\varepsilon_p''$ using Eq. 8-42, we obtain

$$N_0 = T + A_1 A_2 / A_{12}, \qquad (8\text{-}44)$$

where $A_1 A_2 / A_{12}$, for detectors of high intrinsic efficiency, may represent a correction term to T of as small as 5 or 10 per cent.

The two γ-ray peaks of ^{60}Co are not usually completely resolved by a NaI(Tl)-detector system, but $A_1 A_2$ can be computed very closely in terms of $(A_1 + A_2)^2$. This problem does not arise for a radionuclide where the two γ-ray peaks are completely resolved. Such an example is shown on the front cover in the case of ^{88}Y using a very high efficiency detector consisting of two 8-inch NaI(Tl) detectors.

The method of sum-peak coincidence counting is also applied to the calibration of ^{125}I sources in terms of the sum peak formed by the 27-keV K x rays and 35 keV γ ray emitted in coincidence in the decay of that radionuclide. The method was first used by J. S. Eldridge, and P. Crowther in 1964.

The methods described in this section are those that are most often used in a National Standards laboratory to produce standards for the calibration of equipment for the preparation of relative standards. Such calibration standards will be discussed in the next section. In addition, however, to the detector systems needed to effect the calibration of standards, as described above, a standardizing laboratory must also have the equipment that is required for the accurate quantitative dilution of radioactive solutions, and the quantitative preparation of sources, such as dried microgram deposits of solutions on very thin gold-plated collodion films, or liquid scintillators to which solutions have been added. In some cases when isotopically pure standards are required an isotope separator may be used to isolate the radionuclide of interest.

Calibration Standards

All indirect or comparative calibration procedures require some form of prior calibration of the radioactivity measuring system with a calibration standard or standards. Subsequent sample activity measurements are then made using conditions as nearly identical as possible with those used to perform the initial system calibration in order to achieve valid comparative measurements. The required instrument calibration is normally performed by the user, except for certain instruments such as the dose calibrator described in Chapter 7. Such instruments are often supplied precalibrated by the manufacturer with a number of calibration settings for different radionuclides, these may require checking or calibration for settings other than those supplied.

Calibration standards are radioactive sources for which the activity of the principal radionuclide and any measurable radionuclidic impurities are known within specified error limits. Such standards can be procured from either national standardizing laboratories or various commercial suppliers in the form of "certified" calibration standards that are accompanied by a certificate stating the activities and associated errors of all measurable radionuclides present. The certificate itself should conform to the certification guidelines presented in ICRU Report 12.

One problem associated with the use of certified γ-ray-emitting solution standards is that the configuration of the standard (i.e. size, shape and filling of the container) frequently differs from that used by the laboratory for their routine measurements. This problem can, however, be solved by quantitatively transferring the calibration standard, or some fraction thereof, to the desired laboratory container and, when necessary, adding the appropriate inactive carrier solution, which is usually specified in the certificate that accompanies the standard, to achieve the desired volume. This specially prepared standard may then be used to calibrate the measuring system for the required source geometry. Alternatively, if the integrity of the standard is to be preserved, a geometrically similar working standard may be calibrated relative to the certified standard, by means of γ-ray measurements using the same source-detector geometry. The working standard, or standards, can then be used to calibrate instruments for the different laboratory containers in their various geometries.

Once the calibration of a measuring system has been completed, the performance of the system must be routinely monitored at regular intervals, in order to insure the continuing integrity of the calibrations. This is normally done by carrying out periodic measurements with a reference source of a long-lived radionuclide. Such reference sources are normally mounted in rugged containers to prevent damage or leakage and need not be calibrated. Radium-226 in equilibrium with its daughters, or other long-lived radionuclides, such as ^{60}Co or ^{137}Cs-^{137m}Ba, are frequently used for this purpose. The calibration of the instrument should also be checked, usually at longer intervals, by measuring its response to activity standards of the radionuclides that are most frequently used in the laboratory. If these, or the long-lived reference sources, indicate any departure from the normal response the instrument may have to be returned to the manufacturer for repair.

Comparative Measurements of Radioactivity

The general equation for the activity N_0 of a radionuclide source in terms of the response of a radiation counting system, with a nonextendable dead time, is

$$N_0 = \left(\frac{r_{s+b}}{1 - \tau r_{s+b}} - r_b \right) \frac{f(a)f(s)}{\varepsilon G} D, \qquad (8\text{-}45)$$

where r_{s+b} is the count rate due to the source plus background, r_b is the background count rate, τ is the dead time of the system, $f(a)$ is a correction factor for source self-absorption and any other attenuation, $f(s)$ is a correction factor for source and any other scattering, ε is the intrinsic efficiency of the detector, G is the geometrical efficiency, and D is the radioactive-decay correction to a reference time t_0. In all comparative measurements of activity between an unknown and a standard of the same radionuclide, in closely controlled similar geometry, the terms $f(a), f(s), \varepsilon$ and G (and D for longer-lived activities) are the same for both the unknown and known sources.

Therefore, if N_u and N_s are the activities of a source of unknown activity and of a standard source of the same radionuclide, respectively, then

$$N_u/N_s = \left(\frac{r_{s+b}}{1 - \tau r_{s+b}} - r_b \right)_u \Big/ \left(\frac{r_{s+b}}{1 - \tau r_{s+b}} - r_b \right)_s,$$

or $N_u/N_s = R_u/R_s,$

and $N_u = N_s R_u/R_s, \qquad (8\text{-}46)$

where R_u and R_s are the responses of the measuring system, corrected for dead-time losses and background response, in that order. In the case of a direct-current gas-ionization system, τ is equal to zero, there being no dead time in a continuous current.

A comparative, or relative, measurement of the activity of a source of a known radionuclide of unknown activity can be made by one of two general methods.

In the first, the activities of an unknown source and of a calibrated source, or standard, of the same radionuclide are compared, by measuring the relative responses of any kind of suitable detector when one source is substituted for the other in *exactly* the same geometry as possible, *vis-à-vis* the detector. (The source that is not being measured must, of course, be placed in a position where it has no significant effect on the response of the detector to the other source.) The specification "in exactly the same geometry" applies both to the shape or form of the source (i.e. a point source or solutions in standard ampoules) and to its location with respect to the detector. If R_u and R_s are the respective responses of the detector to the unknown and standard sources then the activity of the unknown is given by Eq. 8-46. In such a comparison, the simplest of detectors, such as for example a personnel "chirper" dosimeter can be used to give a rough value of R_u/R_s, just by counting relative numbers of chirps in a given interval of time, when one source is substituted for the other in the same position relative to the dosimeter. If the source is a short-lived radionuclide, decay corrections to a common time t must be made.

The second method of comparative measurement is essentially the same except that the detector, usually in the form of an ionization chamber for the calibration of γ-ray emitting nuclides, is precalibrated for a variety of radionuclides over a range of photon energies. Such is the dose calibrator used now in many nuclear-medicine laboratories, and which is normally calibrated by the manufacturer. The term "dose," used here, has the medical rather than the radiological meaning of the word. The stability of such an instrument should however be checked frequently using sealed sources of long-lived radionuclides, and it should be recalibrated periodically using standards of the radionuclides that are most frequently assayed in the laboratory.

This second method of measurement involves a kind of hidden or overall efficiency which, for lack of a better term, we will call the effective efficiency. The calibration system involves the measurement, in some way or another, of an efficiency which relates response to the activity, or γ-ray emission rate, of a source. In the case of a dose calibrator or ionization chamber the manufacturer or user calibrates it by measuring the read-out or response R_s when a standard of a given radionuclide is placed in the source holder of the instrument. The ratio of response to a given activity of the given radionuclide, R_s/N_s is the efficiency, ε_e, and the activity of any other source of the *same* radionuclide that induces a response R_u is equal to R_u/ε_e. This merely restates Eq. 8-46, but in measuring R_s/N_s and the unknown activity in exactly the same geometry we are eliminating the hidden terms $f(a)$, $f(s)$, ε, and G that comprise the overall, or effective, efficiency ε_e.

If an ionization-chamber system or dose calibrator is precalibrated, the calibration is often given in terms of the so-called K factor which is in reciprocal measure to the overall, effective efficiency ε_e of a detector system. Thus if an ionization-chamber system has a response R_s to an activity N_s of a standard source of a given radionuclide, then K is defined as $K = N_s/R_s$. Therefore, if an unknown source of the same radionuclide gives a response R_u then its activity is given by $N'_u = KR_u$. In such measurements the importance of precise reproducibility of the geometrical and environmental factors involved cannot be too greatly emphasized. Thus both the known and unknown sources, if in the form of solutions, should be placed in containers of the same material having the same shape and dimensions, including thickness of the walls and bases. If for example, the container is fabricated from glass, the barium content of the glass must be carefully controlled. The position of surrounding objects and shielding should also remain unchanged.

The foregoing has been concerned mainly with γ-ray-emitting nuclides, but the same considerations apply to small, thin dried deposits of α- or β-particle-emitting nuclides measured in a windowless, or thin-windowed ionization chamber or, for example, by means of the Lauritsen electroscope depicted in Figs. 6-3 and 6-4.

Both particle-sensitive and γ-ray ionization chambers are normally used in conjunction with sealed radium sources or other long-lived radionuclides. In this case a K factor that is relative to the ionization chamber-system response, R_R, to a long-lived sealed reference source such as ^{226}Ra in equilibrium with its daughters, is used. If R_s is, again, the response to a given calibrated, or standard, source of a radionuclide, activity N_s, then the K factor, relative to the radium response is, $K_R = N_s R_R/R$. If R_u is the response to an unknown source of the same radionuclide measured then, or at some subsequent time, and R'_R is the response at the same time to the same sealed radium source, then the activity of this source of unknown activity is given by $N_u = K_R R_u/R'_R$.

For a sealed pressure gas ionization chamber, such as those containing argon at a pressure of about 20 atmospheres, R_R should not change except for radioactive decay, which in the case of ^{226}Ra is about 0.04 per cent per annum. With an unsealed air chamber, small corrections to the response R_R must be made for changes in the ambient pressure and temperature. If the value of R_R for a pressure chamber, or its corrected value for an unsealed air chamber, changes drastically, then it is necessary to check the system for possible defects.

Discrepancies may also be noted if high-activity sources are measured in ionization chambers, because at high densities of ionization, recombination occurs and the observed ionization current is less than it should be.

In general, simple comparative measurements of activity, using Eq. 8-45 or 8-46, can only be made between sources that are known to contain varying amounts of the *same* radionuclide. Emission rates of α-particle sources can, however, be compared in a simple defined solid-angle counting system in a vacuum, without knowing the identity of the source. This is because G is a constant of the defined solid-angle counter, and ε, f(a) and f(s) are relatively independent of the α-particle energy for thin sources over a range of energies that includes most α-particle-emitting nuclides. It is not generally possible, however, to identify either the principal source of α particles, or any other α-particle-emitting impurity by such simple measurements. The same is also essentially true for defined solid-angle measurements of low-energy photons, such as the x rays emitted by electron-capturing nuclides, the emission rates of which can be measured provided that f(a) is kept to a minimum. For higher-energy photons, however, measured by scintillation or semiconductor detectors, ε, f(a) and f(s) are very dependent on energy, as can be seen from the photon-attenuation coefficients shown in Figs. 3-15 and 3-16. The overall β-particle efficiencies in various β-particle measuring systems are also energy-dependent.

The activity measurements made in nuclear medicine and biology are almost always comparative and are usually restricted to the β-particle emitters ^{3}H and ^{14}C, that are normally assayed by liquid-scintillation counting, or to γ-emitting radiopharmaceuticals that are normally assayed by gas ionization systems, or by scintillation or semiconductor measuring systems. All these systems must, however, be calibrated for response to sources of specified activity of a given radionuclide, or of known γ-emission rate at a given photon energy. The radiations merely serve as a well-specified metrological link between source and detecting system in a given geometrical and environmental condition.

Equation 8-46 contains no terms to allow explicitly for contributions from radionuclides other than the one presumed to be present, although the sources of interest may happen to be mixtures of radionuclides in known proportions, or two or more parent-daughter radionuclides in equilibrium.

If, however, the supplier of a radiopharmaceutical specifies the amounts and identities of radionuclidic impurities present, or if the user is able to identify and quantify them, then the observed response of many comparative radioactivity measuring systems, such as a dose calibrator, can be corrected for contributions from such impurities, provided that the decay modes of each impurity are known.

An alternative and better, but more elaborate approach is to use measuring methods that selectively differentiate between the contributions of the radionuclide of interest and of other radionuclides. Such are the methods of α, β and γ-ray spectrometry. Many α-particle peaks can be resolved and identified by means of the Frisch-grid chamber or surface-barrier detectors with multichannel analyzers. The ranges of α particles and the end-point energies of β-particle spectra can be estimated by absorption measurements (Chapter 3) or liquid-scintillation spectrometry (Chapter 6) and β-particle spectra can be determined by more elaborate β-ray spectrometers. The energies of γ rays can be measured using calibrated scintillation and semiconductor detectors as we learned earlier in this chapter. Half-life measurements can also give an indication of the presence of impurities but often *post de facto*.

In assessing the quantities of radioactive impurities in radioactivity standards or radiopharmaceuticals the most widely used method is that of γ-ray spectrometry. Before discussing these wider uses of γ-ray spectrometry it would be appropriate to pause for a moment to explain the much used terms *integral counting* and *differential counting*.

Integral and Differential Counting. For a detector that is sensitive to the radiations from the radionuclide of interest, the output of which passes into a simple counting system with a lower-level discriminator, all pulses with amplitudes greater than the threshold setting of the discriminator will be recorded. This type of pulse counting is known as integral counting. Suppose, for example, we consider a surface-barrier detector with a monoenergetic α-particle source mounted in the detector vacuum container. If the associated circuitry has a single-channel discriminator that processes pulses of all amplitudes greater than a few milli-volts, then on raising the discriminator threshold from a level corresponding to energies of a few hundred electron-volts up to energies beyond the α-particle energy, an integral count-rate curve would be obtained which would be similar to the upper curve in Fig. 8-11. The sharp drop in count rate occurs as the discriminator threshold energy level passes through the energy corresponding to the monoenergetic α-particle peak, the count rate dropping to zero on the high-energy side of the peak.

If, on the other hand, we use a multichannel analyzer to sort and record counts in intervals of the detector output pulse height corresponding to increments of energy between E and $E + \Delta E$ then the count rate per energy interval ΔE as a function of energy would be as shown in the lower curve of Fig. 8-11. This is similar, in principle, to the γ-ray spectra obtained with a scintillation or semiconductor detector and multichannel analyzer, shown in Fig. 8-3.

On inspection it will be recognized that the lower curve represents the differential of the integral count-rate vs. energy curve. Hence this method of counting is known as differential.

To return to the problem of selectively differentiating between the γ rays of the radio nuclide of interest and unwanted impurities in comparative measurements, a simple integral counting system with a detector that provides an output signal that is proportional to the radiation deposited in it, can only exclude photons that have energies lower than that which corresponds to the discriminator threshold. Thus by setting the discriminator at a threshold voltage corresponding to a given photon energy we can either *exclude* monoenergetic photons emitted by the radionuclide of interest if they be of lower energy, or *include* them if they be of higher energy.

In the case of a simple counting system, or gas ionization chamber, an equivalent means of distinguishing between the photons from the

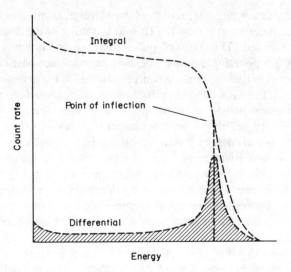

FIG. 8-11   Integral, N(h), and differential, dN/dh, count-rate curves.

radionuclide of interest and those from impurities may be achieved simply by placing absorbers between the source and the detector, in order to attenuate or absorb photons below a given energy. One example of this is the so-called "molybdenum break through measurement," in which a lead absorber is used to remove 90 per cent, or more, of the low-energy ^{99m}Tc γ rays so as to obtain an estimate of the activity of ^{99}Mo that may have broken out of the ^{99}Mo – ^{99m}Tc generator into the injectate. On the other hand, if the detector itself provides an output signal that is proportional to the radiation energy deposited within it, then electronic discrimination can be used to reject the lower-energy ^{99m}Tc photons. Either method would require that the detection efficiency of the system for ^{99}Mo be known for the measuring conditions used, in order to estimate its activity.

Gamma-Ray Spectrometry. By the use of a multiple-discriminator measuring system, such as a multichannel analyzer, in conjunction with a photon detector providing output pulses proportional to the radiation energy deposited within it, pulses corresponding to selected increments of energy, E to $E + \Delta E$, can be recorded to give a *differential spectrum*. In such spectra, γ rays of energies both greater and less than those from the radionuclide of interest are recorded, and their energies can be estimated, and the amounts and identities of impurities possibly established.

This is the basis of γ-ray spectrometry which, used with either NaI(Tl) or Ge(Li) detectors, is now one of the most widely-employed techniques for the identification and quantification of γ-emitting nuclides. Such γ-ray spectra, obtained with mixed radionuclide sources, have already been shown in Fig. 8-3.

From spectra such as those shown in Fig. 8-3 the count rate corresponding to the full-energy peak, or photopeak, of a single γ ray of energy E can be corrected for any contributions from other γ rays emitted by other radionuclides present in the source. This is done by subtracting any background or Compton continuum below the full-energy peak, and then summing the remaining recorded counts for each energy interval, or channel, in the peak; the Compton contribution to the continuum under the peak arises from scattered γ rays of higher energy. Many manual and computerized continuum-subtraction and peak-integration methods have been developed. For guidance in the selection of such methods best suited to a given measurement Section 4 of NCRP Report 58, or the texts by Adams and Dams, or Steyn and Nargolwalla, all referenced in Chapter 6, should be consulted.

When the γ-ray count rate under the photopeak, or full-energy peak, of a particular γ ray, of energy E, emitted in the decay of a given radionuclide in a source has been determined, it is possible to derive the activity of that radionuclide in the source. To do this, it is, however, necessary to know both the photopeak efficiency, ε_γ, at energy E for that source-to-detector configuration and the probability of emission of that γ ray for each disintegration of its parent nucleus. This probability, P_γ, is usually called the *probability per decay* or sometimes, but less correctly, the intensity of that transition.

If $(A_\gamma)_E$ is the count rate under the photopeak of the γ ray of energy E, then the activity of the radionuclide in the source that is emitting these γ rays is

$$N = (A_\gamma)_E / (P_\gamma \varepsilon_\gamma)_E. \tag{8-47}$$

This is, however, a relative measurement of activity in that it is necessary to calibrate the detector either with a suitable standard of the radionuclide in question, or with a series of selected standards to give ε_γ as a function of energy E. A standard should be either one of known γ-ray-emission rate of the γ ray of energy E, or one of known activity and known values of P_γ for the one or more γ rays emitted in its decay. With a series of individual γ-ray-emitting standards, either singly or combined in one source, as illustrated in Fig. 8-3, a calibration curve of ε_γ vs. E can be obtained for a given source-detector configuration and the value of ε_γ for any energy E may be read from the curve. Such a curve is shown in Fig. 8-12. Data that were used to construct this curve were obtained using a standardized mixed γ-ray source of certified γ-ray emission rates for each radionuclidic component, the Ge(Li) spectrum of which is shown in Fig. 8-3. Similar curves of photopeak efficiency as a function of γ-ray energy

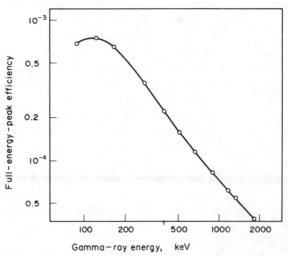

FIG. 8-12   Full-energy vs. energy curve obtained for a 32-cm^3 Ge(Li) detector using a mixed radionuclide point source. A representative spectrum of this source is shown in Fig. 8-3. The source-to-detector distance was 25 cm (from B. M. Coursey, "Use of NBS mixed-radionuclide gamma-ray standards for calibration of Ge(Li) detectors used in the assay of environmental radioactivity", National Bureau of Standards Special Publication 456, 1976).

must also be determined for NaI(Tl)-spectrometer systems to enable photopeak efficiencies for γ rays of different energies to be determined by interpolation. In this way the activities of other radionuclides, for which calibration standards are not available, may be determined by NaI(Tl) or Ge(Li) spectrometry, provided also that the necessary values of P_γ are known.

Similar interpolation techniques can be used for gas ionization chambers and dose calibrators, by plotting $1/K$ or $1/K_R$ as a function of γ-ray energy using available activity standards of monoenergetic γ-ray-emitting nuclides, and then interpolating to obtain the reciprocal of the K factor for radionuclides emitting γ rays of a given energy, but for whic standards are not available. It should be remembered that the K factor $\imath$ the reciprocal of the effective efficiency of the ionization chamber or dose calibrator, and also that the effective K factor for a radionuclide emitting several γ rays of different energies can be calculated from the values o $1/K$, or $1/K_R$, read, at the appropriate energies, from the $1/K$, or $1/K_R$, $vs.$ energy curve. Each $1/K$, or $1/K_R$, is then multiplied by its appropriate P_γ, and the products summed to give an effective reciprocal K factor for that radionuclide. For further details reference may be made to Sections 4.0 and 6.4 of NCRP Report 58.

Choice of Method. The selection of the most suitable detector system for a given radioactivity measurement requires an evaluation of the chemical and physical characteristics of the radioactive source and of the decay parameters of the radionuclide of interest. This evaluation would take into account the type and energies of emitted radiations, the level of activity, the effects of impurities, the geometry of the source, the availability of suitable calibration standards, and the purpose of the measurement itself (i.e. to measure activity, to measure γ-ray-emission rate, to identify individual or constituent radionuclides, and so on).

For sources containing a single radionuclide, integral measuring systems are normally used in order to obtain best sensitivity and precision. For example, gas-proportional- or liquid-scintillation-counting systems are generally selected for comparative activity measurements of α-emitting or β-emitting radionuclides. Simple NaI(Tl) integral counting systems are used for the assay of γ-ray-emitting radionuclides of low to moderate γ-ray-emission rates ($\sim 10^2$ to $10^6\,\mathrm{s}^{-1}$).

For the measurement of high-intensity γ-ray sources and, occasionally for high-activity high-energy β-particle bremsstrahlung sources, a dose calibrator or similar direct-current ionization-chamber measuring system is to be preferred.

To assay sources containing a mixture of γ-ray-emitting radionuclides, γ-ray pulse-height spectrometry is usually the method of choice, both for identification and quantification. The choice between NaI(Tl) and Ge(Li) systems must be contingent upon the degree of resolution required to separate γ rays of nearly equal energies (see, for example, Fig. 8-3).

Resolution and precision are both improved by accumulating as many counts as possible, as the relative standard deviation is approximately equal to $\bar{n}^{-1/2}$. On the other hand, bringing the source close to the detector, to increase the count rate, may cause loss of counts from the photopeaks by summing. Thus it is to be remembered that nearly 4π geometry is the best for the methods of sum-peak coincidence counting (q.v.)

Error. While statistical errors that give rise to imprecision can, as we have seen, be estimated, systematic errors are usually difficult to assess. In comparative measurements systematic errors arise from uncertainties in weighing (in dilution and source preparation), geometry, the value of the calibration standard and its decay parameters, the detection-system response, corrections for dead-time losses and background, and in decay calculations. In order to assess possible systematic errors in any experiment, it is necessary to consider carefully all the possible sources of such error, and then to try to estimate, largely from experience and the sensitivity of the instruments involved, their maximum magnitude. Alternatively an idea of the magnitude of possible systematic error may be obtained by repeating a measurement using completely different and independent measuring procedures.

The overall uncertainty of a measurement is often expressed as the linear sum of the random error, or imprecision, expressed as the standard error (often at the 99 per cent confidence limit), and an estimate of maximum conceivable systematic error.

References

Adams, F. and Dams, R. (1970) See Chapter 6 reference.

Birks, J. B. (1964) See Chapter 6 reference.

Brinkman, G. A. and Aten, A. H. W. (1963) Absolute standardization with a NaI(Tl) crystal—III, *Int. J. Appl. Radiat. Isotopes*, *14*, 503.

Brinkman, G. A. and Aten, A. H. W. (1965) Absolute standardization with a NaI(Tl) crystal—V, *Int. J. Appl. Radiat. Isotopes*, *16*, 177.

Brinkman, G. A., Aten, A. H. W. and Veenboer, J. Th. (1963) Absolute standardization with a NaI(Tl) crystal—I, *Int. J. Appl. Radiat. Isotopes*, *14*, 153.

Brinkman, G. A., Aten, A. H. W. and Veenboer, J. Th. (1963) Absolute standardization with a NaI(Tl) crystal—II, *Int. J. Appl. Radiat. Isotopes*, *14*, 433.

Brinkman, G. A., Aten, A. H. W. and Veenboer, J. Th. (1965) Absolute standardization with a NaI(Tl) crystal—IV, *Int. J. Appl. Radiat. Isotopes*, *16*, 15.

Dixon, W. J. and Massey, F. J. (1969) *Introduction to Statistical Analysis* (McGraw-Hill, New York).

Evans, R. D. (1955) *The Atomic Nucleus*. (McGraw-Hill, New York).

Eldridge, J. S. and Crowther, P. (1964) Absolute determination of ^{125}I in clinical applications, *Nucleonics*, *22* [6], 56.

Fano, U. (1947) Ionization yield of radiations, II. Fluctuations of the number of ions, *Phys. Rev.*, *72*, 26.

Herceg Novi (1973) Proceedings of the first international summer school on radionuclide metrology, *Nucl. Instr. and Meth.*, *112*, 1-398.

Helmer, R. G., Greenwood, R. C. and Gehrke, R. J. (1978) Reevaluation of precise γ-ray energies for calibration of Ge(Li) spectrometers, *Nucl. Instr. and Meth.*, *155*, 189.

ICRU (1968) International Commission on Radiation Units and Measurements. *Certification of Standardized Radioactive Sources*, ICRU Report 12 (International Commission on Radiation Units and Measurements, Washington).

NBS (1963) National Bureau of Standards Handbook 91, *Experimental Statistics*, Mary G. Natrella (U.S. Government Printing Office, Washington, D.C.).

NCRP (1961) National Council on Radiation Protection and Measurements. *A Manual of Radioactivity Procedures*. National Committee on Radiation Protection and Measurements, Report No. 28, published as Nat. Bur. Stand. Handbook 80 (U.S. Government Printing Office, Washington, D.C.).

NCRP (1978) See Chapter 6 reference.

Parratt, L. G. (1961) *Probability and Experimental Errors in Science; an Elementary Survey* (John Wiley, New York).

Quesenberry, C. P. and David, H. A. (1961) Some tests for outliers, *Biometrika*, *48*, 379.

van Roosbroeck, W. (1965) Theory of the yield and Fano factor of electron-hole pairs generated in semiconductors by high-energy particles, *Phys. Rev.*, *139*A, 1702.

Snell, A. H. (Ed.), (1962) See Chapter 6 reference.

Willke, T. A. (1965) Useful alternatives to Chauvenet's rule for rejection of measurement data, National Bureau of Standards Statistical Engineering Laboratory Working Paper W-65-3; also the Proceedings of the 1968 Symposium on Computational Photography.

Zulliger, H. R. and Aitken, D. W. (1970) Fano factor, fact and fallacy, *IEEE Trans. Nucl. Sci.*, NS-17, 187.

General References

Adams, F. and Dams, R. (1970) *Applied Gamma-Ray Spectrometry*, 2nd ed. (Pergamon Press, Oxford and New York).

Birks, J. B. (1964) *The Theory and Practice of Scintillation Counting* (Pergamon Press, Oxford).

Bransome, E. D. (Ed.) (1970) *The Current Status of Liquid Scintillation Counting* (Grune & Stratton, New York).

Dixon, W. J. and Massey, F. J. (1969) *Introduction to Statistical Analysis*, 3rd ed. (McGraw-Hill, New York).

Evans, R. D. (1955) *The Atomic Nucleus* (McGraw-Hill, New York).

Friedlander, G., Kennedy, J. W. and Miller, J. M. (1962) *Nuclear and Radiochemistry*, 2nd ed. (John Wiley & Sons, New York).

Knoll, G. F. (1979) *Radiation Detection and Measurement* (John Wiley & Sons, New York).

Krugers, J. (Ed.) (1973) *Instrumentation in Applied Nuclear Chemistry* (Plenum Press, New York).

National Council on Radiation Protection and Measurements Report No. 58, *A Handbook of Radioactivity Measurements Procedures* (National Council on Radiation Protection and Measurements, Washington, D.C., 1978).

Rutherford, E., Chadwick, J. and Ellis, C. D. (1930) *Radiations from Radioactive Substances* (University Press, Cambridge).

Siegbahn, K. (Ed.) (1965) *Alpha-, Beta-, and Gamma-Ray Spectroscopy* (North Holland Publishing Co., Amsterdam).

Thewlis, T. (Ed.-in-Chief) (1962) *Encyclopaedic Dictionary of Physics* (The MacMillan Company, New York, and Pergamon Press, Oxford and London).

Index of Names

Italic figures refer to Lists of References

Subject Index